Alternative Sources of Protein for Animal Production

PROCEEDINGS OF A SYMPOSIUM

VIRGINIA POLYTECHNIC INSTITUTE
AND STATE UNIVERSITY
Blacksburg, Virginia

July 31, 1972

NATIONAL ACADEMY OF SCIENCES
Washington, D.C.
1973

NOTICE: The project that is the subject of this report was approved by the Governing Board of the National Research Council, acting in behalf of the National Academy of Sciences. Such approval reflects the Board's judgment that the project is of national importance and appropriate with respect to both the purposes and resources of the National Research Council.

The members of the committee selected to undertake this project and prepare this report were chosen for recognized scholarly competence and with due consideration for the balance of disciplines appropriate to the project. Responsibility for the detailed aspects of this report rests with that committee.

Each report issuing from a study committee of the National Research Council is reviewed by an independent group of qualified individuals according to procedures established and monitored by the Report Review Committee of the National Academy of Sciences. Distribution of the report is approved, by the President of the Academy, upon satisfactory completion of the review process.

This study was supported in part by the U.S. Department of Agriculture.

Available from
Printing and Publishing Office, National Academy of Sciences
2101 Constitution Avenue, N.W., Washington, D.C. 20418

LIBRARY OF CONGRESS CATALOGING IN PUBLICATION DATA

Main entry under title.

Alternative sources of protein for animal production.

"[A symposium] held under the joint sponsorship of the Committee on Animal Nutrition of the National Research Council and the American Society of Animal Science at the 1972 meeting of the Society in Blacksburg, Virginia."
Includes bibliographies.
1. Proteins in animal nutrition–Congresses.
I. National Research Council. Committee on Animal Nutrition. II. American Society of Animal Science.
SF98.P7A34 636.08'52 73-5911
ISBN 0-309-02114-6

Printed in the United States of America.

PREFACE

The symposium reported herein was held under the joint sponsorship of the Committee on Animal Nutrition of the National Research Council and the American Society of Animal Science at the 1972 meeting of the society in Blacksburg, Virginia. It sought to examine some of the issues arising from the critical question of protein supplies needed to meet the food needs of the world's expanding human population, in particular, the fact that, in exploiting animal proteins for human consumption, sources of protein for animal production may, in turn, become limiting. In addressing this issue, the symposium explored measures for enhancing protein supplies from known proteinaceous feeds and then identified and evaluated certain new sources.

The organizers and the Committee are grateful to the officers of the society and, in particular, to Dr. O. D. Butler, President, and Dr. T. J. Marlowe, Secretary–Treasurer, who were instrumental in making many of the on-site arrangements. We are grateful, too, to the specialists who appeared on the program and to those of their colleagues and institutional officers who cooperated to make participation posssible.

COMMITTEE ON ANIMAL NUTRITION

Tony J. Cunha, *Chairman*
B. R. Baumgardt
J. Milton Bell
W. H. Hale
John E. Halver
N. L. Jacobson
Robert R. Oltjen
M. L. Sunde
Duane E. Ullrey

ORGANIZING PANEL

J. E. Oldfield, *Chairman*
W. M. Beeson
L. E. Harris

CONTENTS

I
Introduction

J. E. Oldfield

INTRODUCTION

Ever since Berzelius underscored the essentiality of dietary proteins for animals and man, coining the name "protein" from the Greek *proteois,* their nutritional importance has been widely recognized. It is now known that animal proteins contain certain amino acids that make them especially valuable in the human diet. There is much evidence that meat and milk played an important role in the history of man's development and that all strong and stable nations and cultures have emerged in association with animal agriculture (Josephson, 1966). In 1967, the President's Science Advisory Committee on the World Food Problem included the significant comment in its report that, "Only in populations with (an animal protein) class of diet is there no protein malnutrition at present."

In the decade of the 1960's, interest in protein nutrition has intensified, in the light of the widely discussed world food problem. While the details of the issue remain controversial, its general thesis is well known; i.e., a reiteration of the Malthusian doctrine that the increases in population tend to be more rapid than the increases in food production. Malthus' initial warning was not immediately borne out mainly because, it is clear in retrospect, he was unaware of the possibilities of several revolutionary advances: the development of agricultural chemicals and of improved varieties that had the potential for dramatically increasing food production. Those and other means to further improve the world's food resources have not ceased—witness the widely publicized "green revolution"—but concern is voiced by many, including one of the green revolution's principal architects, Norman Borlaug, that it is merely "buying time" against future crises.

If to be forewarned is to be forearmed, then education as to the nature and scope of the world's food crisis is a necessary prelude to successful attack. Probably as eloquent a testimony as any to the importance of protein deficiency is the title of a publication of the Food and Agriculture Organization of the United Nations (FAO), which reads, *Protein: At the Heart of the World Food Problem.* This booklet emphasizes the importance of the qualitative aspects of the problem, noting that there is "no inconsistency in the frequent finding that a country may have a more than adequate total supply of protein, according to a food balance sheet, while at the same time there are many children suffering from the results of a dietary protein deficiency" (FAO, 1967).

The alternatives for supplying protein for the human diet include both animal and plant sources. Arguments as to the relative merits of each seem pointless. Both will be needed, and, while it is true that new plant proteins with greatly enhanced biological value are being synthesized by the plant breeder, it is equally true that the ruminant animal is no less efficient as a synthesizing medium, as demonstrated by Virtanen's experiments (summarized in Virtanen, 1968) and confirmed in this country (Oltjen *et al.,* 1969), in maintaining cattle in good health on a diet devoid of natural protein but provided with nonprotein-nitrogen. Arguments about the relative costs of protein from animal and plant sources and, by implication, their feasibility for impoverished nations will likely continue. In a "strategy statement" for the United Nations, a panel of experts on the protein problem confronting developing countries gave it as their opinion that "Insufficient emphasis has been given to expanding the production of animal proteins in the developing countries" (Subramaniam, 1971). Weight of considered opinion has thus been placed behind the continued production of animal proteins as contributors to dietary protein quality.

If, then, the animal proteins—meat, milk, and eggs—are to contribute significantly to the lessening of man's world food problems, sufficient protein must be allocated to the nutrition of the livestock concerned. To accomplish this is no small task when viewed on a global basis, even though the situation may not yet be critical in certain localized regions, such as the United States and Canada.

Two lines of approach seem promising. Continuing advances in production technology have altered both the quality and the economy of protein supplies from known nitrogen-bearing materials, suggesting that this topic be examined with special reference to cereal grains, forages, oil seeds, and meat by-products. At the same time, significant discoveries are also being made in connection with hitherto unused or little-used protein sources. New sources of protein include a number of different plants, various proteins from the sea, microbial or single-cell proteins, and proteins recycled from animal wastes.

Armed with information on the likely, feasible alternatives for supplying protein in the diets of domestic animals, nutritionists and livestock producers should face the future better prepared to combat existing and expected deficiencies in protein nutrition and, hopefully, to make important contributions toward meeting world food needs.

REFERENCES

Food and Agriculture Organization of the United Nations. 1967. Protein: At the heart of the world food problem. FAO, UN, New York. 62 p.

Josephson, D. V. 1966. General outlook for milk proteins. *In* R. F. Gould [ed.] World protein resources. Adv. Chem. Ser. Am. Chem. Soc. 56:27–36.

Oltjen, R. R., J. Bond, and G. V. Richardson. 1969. Growth and reproductive performance of bulls and heifers fed purified and natural diets. I. Performance from 14 to 189 days of age. J. Anim. Sci. 28:717–722.

Subramaniam, C. 1971. Strategy statement on action to avert the protein crisis in the developing countries. Report of the panel of experts on the protein problem confronting developing countries. UN, New York. 27 p.

President's Science Advisory Committee. 1967. The world food problem. Vol. 2. Report of the Panel on the World Food Supply. The White House. U.S. Government Printing Office, Washington, D.C. 772 p.

Virtanen, A. I. 1968. Some central nutritional problems of the present time. W. O. Atwater Memorial Lecture. ARS, USDA, Washington, D.C. 16 p.

II
Enhancement of Protein Supplies from Known Proteinaceous Feeds

K. J. Frey

IMPROVEMENT OF QUANTITY AND QUALITY OF CEREAL GRAIN PROTEIN

Cereal grains provide the major source of protein and calories for the monogastric livestock populations. Great variation in protein content and quality exist among the cereals, but all tend to have an imbalance of the essential amino acids necessary for full protein utilization. The primary deficient amino acid in all cereal grain proteins is lysine (Mitchell and Block, 1946), but in certain cereals, and for some animals, tryptophan and methionine may be deficient also.

According to Miller (1958), commercial corn samples average only 10.4 percent protein, whereas barley, oats, and rye average over 13 percent (Table 1). For all except young animals, grain that provides 13 percent protein in the ration would be adequate to meet overall nitrogen needs, but because cereal proteins are deficient in certain amino acids, they must be supplemented for monogastric species regardless of their protein level. The protein efficiency ratios (PER) for cereal proteins vary with the crop (Table 2). Corn, wheat, and sorghum have the poorest quality protein, whereas oats, and perhaps rye, have the best. The PER's of the proteins from different cereals seem to be related directly to the proportion of prolamine (alcohol-soluble protein fraction) that the protein contains (Table 3). According to Mossé (1966), this is true because prolamines are very low in lysine.

TABLE 1 Protein Percentages in Commercial Lots of Grains from Cereals[a]

Cereal	Protein Percentage (air-dry basis)		Number of Samples
	Mean	Range	
Barley	13.1	8.5-21.2	1,400
Corn	10.4	7.5-16.9	1,873
Oats	13.3	7.4-23.2	1,850
Rye	13.4	9.0-18.2	112
Sorghum	12.5	8.7-16.8	1,160
Wheat	12.0	8.1-18.5	309

[a]From Miller (1958).

Recently, renewed and expanded research in biochemistry, plant breeding, and animal nutrition has aimed at making cereals better sources of protein for livestock (and human) nutrition. I will try to describe for each of the cereals used in livestock rations the events that have led to this expanded research, the progress to date, and the probability of success. Each of the cereals—corn, barley, wheat, oats, sorghum, and *Triticale*—will be discussed, and, finally, I will discuss the negative yield–protein content correlation so often reported.

CORN

The earliest verifiable studies on the protein of corn (*Zea mays*) were done by Gorman in the early nineteenth century. He reported that corn grain had 3.3 percent alcohol-soluble protein (zein) that constituted two fifths of the total protein. Extensive research, however,

TABLE 2 PER of Cereal Proteins in Diets of Rats

Protein Source	Protein Percentage in Ration			
	7.5[a]	9.5[a]	9.0–10.0[b]	8.0–10.0[c]
Barley	1.7	–	1.6	2.0
Corn	1.6	–	1.4	–
Oats	2.1	2.5	1.8	2.2
Rye	2.2	1.8	1.3	–
Sorghum	–	–	0.7	–
Wheat	1.4	1.7	0.9	1.7

[a]From Jones *et al.* (1948).
[b]From Howe *et al.* (1965).
[c]From Morkuze (1937).

TABLE 3 Percentages of Prolamine Fractions in Grain Proteins of Various Cereals[a]

Crop	Prolamine Name	Percent of Protein
Barley	Hordein	40
Corn	Zein	50
Oats	Avenin	12
Rye	Secalin	40
Sorghum	Kafarin	60
Wheat	Gliadin	45

[a]From Mossé (1966).

on corn protein did not get under way until the twentieth century. The low quality of corn protein (Table 4) is due to deficiencies in lysine and tryptophan (Marias and Smuts, 1940). Zein is the site of deficiencies of these amino acids (Osborne and Clapp, 1908; Osborne and Mendel, 1914): In fact, in experiments conducted by Willcock

TABLE 4 Percentages of Amino Acids in Grain Protein from Normal, *Opaque-2*, and *Floury-2* Endosperms[a]

Amino Acid	Endosperm Type (g/100 g protein)		
	Normal	*Opaque-2*	*Floury-2*
Alanine	10.0	7.2	8.0
Arginine	3.4	5.2	4.5
Aspartic acid	7.0	10.8	9.0
Cystine	1.8	0.9[b]	1.8
Glutamic acid	26.0	19.8	19.0
Glycine	3.0	4.7	3.4
Histidine	2.9	3.2	2.2
Isoleucine	4.5	3.9	4.0
Leucine	18.8	11.6	13.3
Lysine	1.6	3.7	3.3
Methionine	2.0	1.8	3.2
Phenylalanine	6.5	4.9	5.1
Proline	8.6	8.6	8.3
Serine	5.6	4.8	4.8
Threonine	3.5	3.7	3.3
Tryptophan	0.3	0.7	0.8
Tyrosine	5.3	3.9	4.5
Valine	5.4	5.3	5.2
Protein[c]	12.7	11.1	13.6

[a]From Nelson (1969).
[b]In other samples, value for cystine was 1.8.
[c]Percent of grain.

and Hopkins (1913), mice fed zein as the sole protein source in a ration died in a short time.

In 1896, a selection experiment was begun at the Illinois Agricultural Experiment Station to produce high- and low-protein strains of corn (Hopkins, 1899), and after 60 generations of selection, the protein content had been modified from 10.9 percent in the original Burr's dent corn to 22.8 and 4.5 percent in the high- and low-protein strains, respectively (Leng, 1961). These selection experiments are of significance to animal nutrition research because they furnish the material for much later experimentation. Showalter and Carr (1922) showed that protein quality of corn decreased as the protein percentage in the grain increased because a disproportionate part of the increase was made up of zein. Hansen *et al.* (1946), using corns with a range of kernel protein from 6.0 to 19.0 percent, found that total protein increased only half as fast as zein. Thus, in low-protein corn the zein:protein ratio was 0.3, whereas in high-protein corn it was 0.5. In a series of corn populations, Frey (1951a) found that zein increased from 40 to 95 percent faster than total protein. From 64 corn inbreds analyzed, Frey (1948) found none that deviated significantly from the standard pattern of zein:protein relationship, and in a selection experiment, Frey *et al.* (1949) made no progress in selecting for a lower zein:protein ratio.

To summarize, despite extensive work done on corn protein and the attempts that were made to improve its nutritional value, no progress had been possible up to that time. Normal corn protein in the endosperm of the kernel invariably had a high proportion of zein, a protein fraction nearly devoid of tryptophan and lysine. Furthermore, as corn protein content in the grain was increased by selection, the protein, gram for gram, became lower in quality.

In the early 1960's, a break came in the search for better nutritional quality in corn grain when Mertz *et al.* (1964) and Nelson *et al.* (1965) reported that two endosperm mutants of corn called *opaque-2* and *floury-2* significantly modified the amino acid composition of corn protein (Table 4). The *opaque-2* gene increases lysine, tryptophan, arginine, and aspartic acid contents and decreases glutamic acid, alanine, and leucine in corn endosperm protein. The *floury-2* gene causes approximately the same amino acid changes and, additionally, increases the methionine content. Nelson (1969) investigated the amino acid composition of embryo, endosperm, leaf, and pollen tissues from *opaque-2* and normal isolines. The only tissue with dissimilar amino acid compositions was the endosperm.

TABLE 5 Protein Percentages in Corn Grain of Normal, *Floury-2,* and *Opaque-2* Genotypes as Reported by Several Researchers

Percent Protein in			
Normal	*Opaque-2*	*Floury-2*	Reference
12.7	11.1	13.6	Nelson *et al.* (1965)
10.7	–	15.0	Nelson (1969)
10.8	11.1	–	Alexander *et al.* (1969)
11.2	12.9	11.6	Sreeramulu *et al.* (1970)

Whether *opaque-2* and *floury-2* cause changes in total protein content of the corn grain is controversial (Table 5). Nelson (1969) reported a substantial increase of 4.3 percent due to *floury-2,* but no other researcher has been able to substantiate his work. According to Alexander (personal communication from D. E. Alexander, University of Illinois, Urbana, Illinois), the protein content modification caused by either *opaque-2* or *floury-2,* if any, probably depends upon interaction of the mutants with the remainder of the genotype.

When first discovered, Mertz *et al.* (1964) ascribed the increased lysine of *opaque-2* corn to a reduced zein content. Whereas normal corn protein contains 40 percent zein, *opaque-2* corn protein contains only 15 percent (Murphy and Dalby, 1971). Recently, however, it has been shown that the *opaque-2* gene changes both the proportions of the protein fractions and the lysine contents within the various proteins (Table 6). There is a sharp reduction in the ethanol-soluble protein and concomitant increases in the salt- and alkali-soluble proteins. No major change occurs in the lysine percent in the

TABLE 6 Fractional Distribution of Protein and Percents of Lysine in Various Protein Fractions in *Opaque-2* and Normal Corn Grain[a]

	Fraction of Total Protein (%)		Lysine Content (%)	
Protein Fraction	*Opaque-2*	Normal	*Opaque-2*	Normal
Saline solution	21	6	2.1	1.8
Butanol solution	2	7	0.4	0.2
Ethanol solution	16	33	0.4	0.3
Alkali solution	39	27	4.1	1.7
Residue	18	22	6.2	2.8

[a]From Murphy and Dalby (1971).

TABLE 7 PER for Proteins from Several Sources[a]

Protein Source	PER
Casein	3.1
Opaque-2 corn	2.8
Soybeans	2.7
Normal corn	1.6

[a]From Mertz *et al.* (1965, 1966).

saline-, butanol-, and ethanol-soluble fractions, but the lysine percent in the alkali-soluble and residue protein fractions is more than doubled.

Fuchs (1970) postulated that it would be possible to select modifier genes in standard corns that would affect the high-lysine expression of *opaque-2,* but Feist and Lambert (1970) could find no evidence for such modifier genes in a backcrossing program to convert six inbreds.

Mertz *et al.* (1965, 1966), using rats as test animals, found *opaque-2* corn protein had a PER value equivalent to soybean protein and nearly equal to casein (Table 7). The PER's for normal and *opaque-2* corn were 1.6 and 2.8, respectively. Rats fed normal and *opaque-2* corn as protein sources for 4 weeks gained 25 and 100 g, respectively, of body weight.

Opaque-2 corn has been tested extensively in pig and chicken rations. Beeson *et al.* (1966) and Cromwell *et al.* (1967, 1969) showed *opaque-2* corn to be nutritionally superior to and more digestible than normal corn for swine. An *opaque-2* ration supported swine growth as well as a ration with normal corn and soybeans. For 3-week-old chicks that require a ration with 20 percent protein, Fonseca *et al.* (1970) found *opaque-2* promoted more rapid growth than did normal corn, but lysine supplementation was still required for maximum growth rate. Fonseca *et al.* (1969) reported that egg production percentage of layers increased from 71 percent, when a normal corn ration was fed, to 78 percent when *opaque-2* was the protein source.

Both chemical analyses and feeding tests have established the superiority in feeding quality of protein from *opaque-2* and *floury-2* corn over that of normal corn. The superiority is due to increased contents of lysine and tryptophan in *opaque-2* and lysine, tryptophan, and methionine in *floury-2* protein. Basically, the increased

TABLE 8 Agronomic Production and Performance of *Opaque-2* and Normal Three-Way Corn Hybrids Produced with Bc_1-Derived Lines[a]

Endosperm Type	Yield (q/ha)	Moisture (%)	Erect Plants (%)	Weight/100 Seeds (g)	Cracked Kernels (%)
Normal	64	21.0	71	30.4	3.8
Opaque-2	59	25.2	71	29.0	7.2
Difference	-5[b]	+4.2[b]	0	-1.4[b]	+3.4[b]

[a]Alexander *et al.* (1969).
[b]Significant difference at 5 percent level.

contents of these amino acids are due to changes in the proportions of several protein fractions and changes in amino acid composition of at least two protein fractions.

Since the *opaque-2* and *floury-2* traits are monogenically inherited (Emerson *et al.*, 1935), inbred corn lines can be converted to either endosperm mutant quickly and easily via backcrossing. Both mutants, however, cause associated deleterious effects on other plant traits. For example, Alexander *et al.* (1969) found that when *opaque-2* and normal counterpart hybrids (using Bc_1-derived lines) were compared, grain yield and weight per 100 seeds averaged 8 and 5 percent lower, respectively, in *opaque-2* hybrids, and moisture content and percent of cracked kernels at harvest were higher (Table 8). In comparisons of hybrids from Bc_2- and Bc_4-derived inbred lines, the yields averaged lower in *opaque-2* hybrids by 15 and 8 percent, respectively (Table 9). Sreeramulu and Bauman (1970) and Sreeramulu *et al.* (1970) also report that *opaque-2* and *floury-2* cause significantly reduced grain yields and kernel weights. Lambert *et al.* (1969) reported an average yield reduction of 8 percent for three-way crosses of *opaque-2* com-

TABLE 9 Agronomic Production and Performance of *Opaque-2* and Normal Single Crosses of Corn Produced with Bc_2- and Bc_4-Derived Lines[a]

Endosperm	Yield (q/ha)		Moisture (%)		Weight/100 Seeds (g)
Type	Bc_2	Bc_4	Bc_2	Bc_4	Bc_2
Normal	78	58	24.0	22.0	34.6
Opaque-2	66	54	26.4	26.0	29.9
Difference	-12[b]	-4[b]	+2.4[b]	+4.0[b]	-4.7[b]

[a]From Alexander *et al.* (1969).
[b]Significant difference at 5 percent level.

TABLE 10 Grain Yields (q/ha) from Three-Way *Opaque-2* and Normal Corn Hybrids[a]

Single-Cross Tester		Inbred							
Designation	Yield	B65	C121E	C144	H49	M017	Oh7N	R109B	Va36
R801 × R802 (Nor)	61	73	66	72	73	76	71	61	66
R801 × R802 (*Op-2*)	59	65	65	62	68	52	67	60	56
R802 × R803 (Nor)	69	54	55	61	70	56	64	60	51
R802 × R803 (*Op-2*)	59	61	60	56	53	48	59	62	54

[a]From Lambert *et al.* (1969).

pared to normal counterparts. Two *opaque-2* hybrids (R 802 X R 803) X B 65 and (R 802 X 803) X C 121E, yielded more than their normal counterparts, and in six other pairs the yields were not significantly different (Table 10), but note that the pairs, in which yields of *opaque-2* hybrids were equivalent or superior to their normal counterparts, tended to be those where the yield of the normal hybrid was low. In Illinois in 1970, 30 *opaque-2* single-cross hybrids yielded from 47 to 98 q/ha, whereas five commercial hybrids yielded from 51 to 102 q/ha, and in 1971, comparable ranges were 44 to 88 and 45 to 92, respectively (personal communication from D. E. Alexander, University of Illinois, Urbana, Illinois). The grain moisture content at harvest and lodging resistance of the *opaque-2* hybrids were comparable to the checks.

In summary, the *opaque-2,* and perhaps the *floury-2,* endosperm types significantly improve the feeding value of corn grain protein. Either mutant can be transferred easily to existing inbreds, but *opaque-2* hybrids have tended to have inferior yields and kernel quality and a slower drying rate at harvest. Some recent *opaque-2* hybrids seem to be as productive as their normal endosperm counterparts. In fact, Harpstead (1971) reported the release of *opaque-2* hybrids for commercial use in South America, and at least one U.S. corn seed company has *opaque-2* hybrids for sale. Modifier genes have been found that make the *opaque-2* seeds as dense as normal seeds without affecting the lysine content. The double-mutant hybrid, i.e., *opaque-2* and *floury-2* in the same hybrid, has not been promising, but the search is continuing for additional mutants that modify the amino acid composition of corn endosperm protein, and McWhirter (1971) has reported that *opaque-2* causes both high lysine and high methionine in corn protein.

BARLEY

Somewhat over half of the barley (*Hordeum vulgare* and *H. distichum*) grain produced in the United States is used as livestock feed. The biological value of barley protein is only intermediate among the cereals (Table 2) because it has deficiencies—first, in lysine and, second, in threonine (Howe *et al.*, 1965). As with other cereals, the low lysine content in barley protein is due to the virtual absence of lysine in hordein, the alcohol-soluble protein fraction (Folkes and Yemm, 1956). Bishop (1928) reported that hordein increases as the protein content of barley grain goes up, and Munck *et al.* (1969) found correlations between protein content in the grain and lysine percentage in the protein of −0.89** and −0.50* for 16 commercial cultivars and 18 experimental lines, respectively. Viuf (1969) tested 650 barley cultivars and found that cultivars with 9.5–10.9 percent protein had 4.2 percent lysine in the protein, whereas those with 17.8–18.1 percent protein had only 2.7–3.1 percent lysine (Table 11).

TABLE 11 Percentages of Protein in Grain and Lysine in Protein for Barley Cultivars[a]

Percent Protein	Percent Lysine in Protein
9.5	4.2
10.9	4.2
15.6	3.5
16.3	3.2
17.8	3.1
18.1	2.7

[a]From Viuf (1969).

After Mertz *et al.* (1964) reported the marked effect the *opaque-2* gene had in increasing the lysine content in corn grass protein, Hagberg and Karlsson (1969) tested over 1,000 lines from the Barley World Collection for grain protein content and dye-binding capacity. The

**Significant at the 1 percent level.
*Significant at the 5 percent level.

TABLE 12 Percentages of Protein in Grain and Lysine in Protein for C.I. 3947 Barley Strain and Other Cultivars[a]

		Percent of	
Barley Lines	Number	Protein	Lysine
Svalof, Sweden–1967			
Foma lines	7	11.0	3.4
C.I. 3947	1	16.9	4.0
Svalof, Sweden–1968			
Low-protein lines	3	10.7	3.9
High-protein lines	5	15.0	3.4
C.I. 3947	1	14.5	4.2
Aberdeen, Idaho–1966			
Commercial cultivars	4	12.3	3.6
C.I. 3947	1	18.4	4.4

[a]From Munck *et al.* (1969).

latter is a measure of basic amino acids in the grain, and since lysine is a major basic amino acid, this is a practical test for lysine content. They found a correlation of –0.93 between protein content and dye-binding capacity for all lines, but one naked strain, C.I. 3947 from Ethiopia, had both high-protein percentage (17.0) and high-lysine in the protein (4.2). This is approximately the lysine content normally found in barley grain with 10 percent protein. Munck *et al.* (1969), from extensive studies of the composition of C.I. 3947 and other cultivars grown in the same experiments (Table 12), confirmed the high-lysine content of C.I. 3947 across several environments.

The high-lysine–high-protein trait has been named "*hiproly.*" It causes a marked reduction in hordein content of barley grain, but not as marked reduction as the *opaque-2* trait causes in corn. According to Munck (1971), *hiproly* causes increases in lysine, aspartic acid, and methionine and decreases in glutamic acid, cystine, and proline (Table 13).

Hiproly is conditioned by a single gene at a locus on chromosome 7 of the barley genome, and it has been given the symbol *lys.* All segregates with the *hiproly* trait show the modified protein and amino acid compositions.

In nutrition experiments utilizing rats and mice, lines from crosses of *hiproly* X normal barley cultivars confirm the improved nutritional quality of the high lysine types. The net protein utilization values for normal and *hiproly* lines were 60 and 69 percent, respectively, and the biological values were 71 and 81, respectively, in rations with

TABLE 13 Percentages of Protein in Grain and of Amino Acids in Protein of High- and Low-Protein and *Hiproly* Barley Cultivars[a]

	Barley Lines[b]		
Constituent	Low Protein	High Protein	*Hiproly*
Protein	10.7	15.0	14.5
Aspartic acid	6.3	5.5	6.7
Cystine	2.3	1.9	1.7
Glutamic acid	25.3	26.9	23.8
Lysine	3.9	3.4	4.2
Methionine	1.5	1.4	1.7
Proline	10.2	11.7	11.1

[a]From Munck (1971).
[b]Protein expressed as percent of grain and amino acids expressed as percent of protein.

barley as the only protein source at a level of 9.4 percent.

C.I. 3947, the original barley line with the *hiproly* trait, is poorly adapted to the temperate zone and produced naked seeds. Munck (1971) reported that *hiproly* and normal lines from crosses of C.I. 3947 with commercial barley cultivars are equivalent in daylight response, plant height, number of spikes per plant, and number of seeds per plant. Generally, *hiproly* lines have lower kernel weight than the normal lines, but a few had normal kernel weight. To date no yield results have been reported for *hiproly* segregates.

Obviously, the *hiproly* trait would be desirable in feed barley cultivars. It would enhance the protein and lysine contents of the barley grain, but the content of the second limiting amino acid, threonine (Howe *et al.*, 1965), would not be improved. The monogenic inheritance of *hiproly* makes it easy to transfer to already good barley cultivars via backcrossing. So, if experimentation shows that this trait causes no significant associated deleterious effects on grain yield and other agronomic traits, it is likely that *hiproly*-type barley cultivars will be available before the end of this decade.

WHEAT

Wheat (*Triticum durum* and *T. aestivum*) is predominantly a human food, but it is used enough for livestock feed that we should consider attempts to improve this crop as a protein source. The biological value for wheat protein tends to be similar to that for corn, and it is deficient in lysine (Hegsted *et al.*, 1954) and, perhaps, tryptophan (Csonka,

1937) and methionine (Baumgarten *et al.*, 1946). However, since lysine is the primary limiting amino acid, plant breeders have given most attention to searching for genes for high lysine. Johnson *et al.* (1972a) analyzed 15,000 lines of bread and durum wheats from the Wheat World Collection, but to date, they have found nothing comparable to the high-lysine content produced by *opaque-2* and *floury-2* in corn or *hiproly* in barley. The lysine content has varied from 2 to 4 percent of protein, but lysine in protein of wheat decreases as the protein percentage increases in the grain. Villegas *et al.* (1970) reported correlations between protein percentage in grain and lysine content in protein were -0.68 for hard red spring and -0.77 for durum wheat. Lawrence *et al.* (1958) also reported a correlation of -0.73, but this negative relationship existed only for grain with protein percentage below 13.5. However, Johnson *et al.* (1971) did find one line, Nap Hal, that has high-protein percentage and one-sixth more lysine in the protein than other wheats of comparable protein contents (Table 14). However, this difference is so small (0.5 percent lysine in protein) that only time can prove if it has practical value.

Lawrence *et al.* (1958) analyzed the grain of wild wheat species for protein and lysine and found *T. pyramidale* and *T. sphaerococum,* both with 15 to 20 percent protein, had 3.5 percent lysine in their proteins. Several *Agropyron* species (*A. amurense, A. trachycaulum,* and *A. riparium*) also have high-lysine content in high protein samples. Villegas *et al.* (1970) reported five samples of *T. boeticum* had mean values of 19.1 percent protein and 2.9 percent lysine in protein. For comparison purposes, Johnson *et al.* (1972b) found 47 bread wheat cultivars with a mean of 19.0 percent pro-

TABLE 14 Percentages of Protein and of Three Amino Acids in Protein in Normal (Aniversario) and High Lysine (Nap Hal) Wheat Cultivars[a]

	Cultivar	
Constituent	Aniversario	Nap Hal
Protein[b]	20.7	21.3
Lysine[c]	2.36	2.81
Methionine[c]	1.32	1.57
Threonine[c]	2.88	3.18

[a] From Johnson *et al.* (1971a).
[b] Percent of grain.
[c] Percent of protein.

TABLE 15 Grain Yields and Grain Protein Percentages of Experimental Lines and Cultivars of Wheat[a]

Cross or Cultivar	Selection Number	Grain Yield (q/ha)	Protein (%)
Atlas 66 × Wichita	631068	26.8	18.5
	631423	30.2	17.5
	631417	25.0	17.5
	631168	28.6	17.1
	631250	26.6	17.0
Lancer[b]	–	27.2	14.8
Comanche[b]	–	19.1	14.4
Wichita[b]	–	19.0	14.0

[a]From Johnson *et al.* (1969).
[b]Commercial cultivars.

tein had a mean of 2.8 percent lysine in protein. Even though some wild wheat species have the combination of high protein and high lysine, it is not likely that this combination can be transferred to cultivated wheat soon, if ever.

Extensive work has been done at the Nebraska Agricultural Experiment Station simply to increase the protein content of wheat grain (Johnson *et al.*, 1963, 1967, 1971, 1972a,b; Mattern *et al.*, 1968; Stuber *et al.*, 1962). Atlas 66, a soft winter wheat reported by Middleton *et al.* (1954) to have high-protein content, has been crossed with hard red winter wheat cultivars—and some lines from the cross Atlas 66 × Wichita yield as well as Lancer, a current commercial cultivar—and have an increase of 2.0 to 2.5 percent in protein (Table 15). The distribution of essential amino acids in Atlas 66, Comanche (a hard red spring wheat with lower protein content), and several high-protein selections from Atlas 66 × Comanche (Table 16) were shown to be similar by Mattern *et al.* (1968). An interesting feature of these high-protein Nebraska lines is that they do not extract excessive nitrogen from the soil, but they are more efficient in translocating the nitrogen from vegetative tissue to the grain.

With wheat, lacking genes that significantly alter the amino acid composition of the grain protein, plant breeders chose to work toward increasing protein percenge in the grain. As a result, experimental lines of hard red winter wheat with a one-fifth higher protein content have been achieved. Additional increases in protein content of wheat grain will likely be more difficult to achieve.

TABLE 16 Percentages of Protein in Grain and of Essential Amino Acids in Protein of Wheat Cultivars and Lines[a]

	Cultivar or Line				
Amino Acid	Atlas 66	Comanche	2499	2500	2509
Isoleucine	3.8	3.9	3.8	3.8	3.9
Leucine	7.6	7.5	6.8	7.4	7.7
Lysine	3.3	3.2	3.3	3.2	3.5
Methionine	1.1	1.7	1.6	1.7	1.8
Phenylalanine	5.3	5.6	5.9	5.4	5.5
Threonine	3.4	3.5	3.1	3.2	3.3
Tyrosine	3.8	3.9	4.0	3.8	3.9
Valine	4.6	4.6	4.6	4.5	4.7
Protein	18.0	15.0	18.2	18.3	18.3

[a]From Mattern *et al.* (1968).

SORGHUM

Sorghum (*Sorghum bicolor*) grain protein is among the poorest of the cereal grains in biological value (Table 2), and its primary amino acid deficiency is in lysine (Pond *et al.*, 1958; Vavich *et al.*, 1959). The correlation between protein percentage in sorghum grain and lysine percentage in the protein was found to be –0.34 by Collins and Pickett (1972) and Deosthale *et al.* (1970) and –0.33 by Virupaksha and Sastry (1968). As would be expected, Vavich *et al.* (1959) and Waggle *et al.* (1966) confirmed that, when fed in isonitrogenous rations, the sorghum protein from low-protein strains promotes growth of chicks and rats faster than that from high-protein strains. For chicks, tyrosine and phenylalanine also may be deficient in sorghum protein (Deyoe and Shellenberger, 1965).

The possibility of improving sorghum as a protein source, of course, will depend upon the genetic variation available in the species. Stephenson *et al.* (1971) found large differences in amino acid compositions of sorghum lines and even greater differences in amino acid availability for chicks. Presumably, these differences were genetic in origin. Deyoe and Shellenberger (1965) found large ranges of amino acids among the 30 sorghum lines they analyzed (Table 17). Collins and Pickett (1972), from analyzing sorghum hybrids, found protein varied from 11.3 to 15.5 percent of the grain and lysine varied from 1.3 to 2.0 percent of protein. The inheritance of all amino acids in

TABLE 17 Means and Ranges of Amino Acid Percentages in Proteins from 30 Sorghum Lines[a]

Amino Acid	Percentage in Protein	
	Mean	Range
Alanine	9.2	7.3–10.7
Arginine	2.7	2.1– 3.4
Aspartic acid	6.3	4.8– 7.7
Cystine	1.0	0.5– 1.4
Glutamic acid	21.1	17.0–24.9
Glycine	3.0	2.4– 3.5
Histidine	2.1	1.7– 2.3
Isoleucine	3.8	2.9– 4.8
Leucine	13.1	10.2–15.4
Lysine	2.0	1.6– 2.6
Methionine	1.3	0.8– 2.0
Phenylalanine	4.8	3.8– 5.5
Proline	7.7	6.0– 8.9
Serine	4.1	3.2– 5.5
Threonine	3.0	2.4– 3.7
Tyrosine	1.6	1.2– 2.5
Valine	4.9	4.0– 5.8
Protein[b]	11.0	–

[a] From Deyoe and Shellenberger (1965).
[b] Percent of grain.

sorghum was quantitative and additive in a study by Haikerwal and Mathieson (1971).

The most promising lead to date for improving sorghum as a protein source has come from Virupaksha and Sastry (1968), who found one sorghum line, 160 Cernum that had high protein (17.7 percent) and high lysine in the protein (2.1 percent). Apparently, this protein–lysine relationship was caused by 160 Cernum having a low-prolamine and high-glutelin content. However, the protein from this sorghum line would be judge deficient in lysine when compared to that from *opaque-2* corn or oats, both of which have more than 4.0 percent lysine in their protein. Deosthale *et al.* (1970) also reported four sorghum lines that are classed as high protein and high lysine.

It appears that no genetic type of sorghum has been found that has the profound effect on amino acid composition of grain as that caused by *opaque-2* in corn. Probably, protein content of sorghum grain could be raised, but data available to date predict that the quality of such protein would be lowered.

OATS

Of the cereals used as livestock feed, oats (*Avena sativa* and *A. byzantina*) and rye (*Secale cereale*) appear to have protein with the best biological value (Table 2). Hischke *et al.* (1968) found that proteins (60 percent oats by weight in the rations) from seven oat cultivars all had similar PER's of 2.3 and 2.4 when fed to rats, but Weber *et al.* (1957) showed that Cimarron, Forkedeer, and Winter Fulghum cultivars produced high body-weight gains when fed to rats, but De Soto and Selection 4829 produced low gains. There were no associations between lysine or methionine contents of the cultivars and the gains these cultivars produced.

The primary limiting amino acid in oat protein, as for all other cereals, is lysine, and the alcohol-soluble protein fraction, avenin, is nearly devoid of this essential amino acid (Frey, unpublished data). However, total protein from oat flour, has relatively high levels of both lysine and tryptophan when compared to the other feed grains (Table 18). Oat protein contains 4.5 percent lysine whereas the protein from other cereals vary from 2.0 percent for wheat to 3.5 percent for corn. The second and third limiting amino acids in oat-grain protein are threonine and methionine (Robbins *et al.*, 1971). Furthermore—contrary to the situation with corn, barley, and wheat—the protein composition data suggest that the biological value of oat protein does not deteriorate as the protein percentage in the grain increases. Frey (1951b) found that oat cultivars with a range of protein

TABLE 18 Percentages of the Essential Amino Acids (Anhydro) in Flour Proteins From Five Cereals[a]

Amino Acid	Percentage of Protein in				
	Wheat	Rye	Barley	Oats	Corn
Isoleucine	3.6	3.6	3.6	3.8	3.6
Leucine	6.7	6.7	7.2	7.7	11.6
Lysine	2.0	3.2	3.1	4.5	3.5
Methionine	1.3	1.7	1.7	1.8	2.0
Phenylalanine	5.1	4.9	5.5	5.2	4.9
Threonine	2.7	3.4	3.3	3.7	3.9
Tryptophan	1.1	1.8	2.0	2.0	0.9
Tyrosine	2.6	2.1	2.7	2.6	2.3
Valine	3.7	4.4	4.6	5.0	4.9

[a]From Ewart (1967).

TABLE 19 Percentages of Protein and Avenin and Ratios of Avenin:Protein for Grain of Selected Oat Varieities[a]

Cultivar	Protein (%)	Avenin (%)	Avenin:Protein
Huron	9.3	1.8	0.19
Colo	10.7	2.0	0.19
Wolverine	11.9	2.1	0.18
Beaver	12.6	2.3	0.18
Bonda	13.4	2.4	0.18
C 3656	14.5	2.6	0.18
C 5298	15.8	3.0	0.19

[a]From Frey (1951b).

from 9.3 to 15.8 percent all had avenin:protein ratios of 0.18 or 0.19 (Table 19). Furthermore, Robbins *et al.* (1971) found the correlation of grain protein percentage and lysine percentage in the protein was very small. Recent analyses on lines from *Avena sterilis*, a wild oat from Israel, show that the amino acid percentages in the grain protein from this species also remains constant over a wide range of protein percentages (Table 20). Because oat protein has a good biological value (relative to other cereals) and its amino acid composition is highly constant over a wide range of protein percentages, plant breeders have decided to concentrate their efforts on increasing protein percentage in the grain.

TABLE 20 Percentages of Protein and Amino Acids in Grains of *Avena Sterilis* Lines[a]

	Line Number		
Constituent	1	2	3
Protein[b]	17.0	25.1	21.7
Cystine	1.8	1.7	1.6
Leucine	7.8	7.8	7.9
Lysine[c]	4.0	3.9	4.1
Phenylalanine	5.5	5.5	5.7
Threonine	3.3	3.4	3.4
Tyrosine	3.3	3.4	3.4
Valine	5.7	5.7	5.9

[a]From D. E. Western, Quaker Oats Co., Chicago, Ill. (unpublished data).
[b]Expressed as percent of dry weight.
[c]Amino acids are expressed as percent of protein.

Whole grain of oats generally has from 9 to 16 percent protein (Frey and Watson, 1950), whereas the maximum protein in groats (naked caryopses) of commercial cultivars Briggle (1971) has been quoted as 20 percent. Three general approaches are being used currently to increase the protein percentage of future oat cultivars: (a) recombination of protein genes from various cultivars and lines of *A. sativa* and *A. byzantina* (cultivated oats); (b) artificial induction of mutations that cause high-protein percentage; and (c) transfer of high-protein genes from *A. sterilis* to *A. sativa* cultivars.

Robbins *et al.* (1971) analyzed groats from 289 oat lines extracted from the Oat World Collection and found a range of groat-protein percentages from 12.4 to 24.4 and a mean of 17.1. Only Florad (24.4 percent) and Yancey (24.2 percent) exceeded the mean by three standard deviations. Frey (unpublished data) analyzed nearly 3,300 lines from the same Oat World Collection and found a range from 7.8 to 21.9 percent protein in the whole grain. The data collected by Robbins *et al.* (1971) and Frey suggested that it should be possible to elevate the groat-protein percentages of future oat cultivars above the 15.0 to 18.0 percent of current cultivars by recombination of high-protein genes already present in cultivated lines. In a recent study at Iowa State, we analyzed groats from 192 random F_9-derived oat lines from a bulk population that was constructed to provide recombination of genes for lodging resistance. The range of protein percentages was from 14.6 to 32.3 (Table 21). From the same ex-

TABLE 21 Frequency Distribution for Groat-Protein Percentages of F_9-Derived Oat Lines from a Bulk Population[a]

Protein Percentage Class Center	Number of Lines
33	1
31	2
29	5
27	2
25	4
23	27
21	46
19	71
17	29
15	5
MEAN	20

[a]Unpublished data.

perimental area, commercial cultivars had groat-protein percentages from 15.8 to 19.7. In a more extensive study of 1,000 F_9-derived lines from another bulk population, the groat-protein percentages ranged from 11.8 to 23.2, whereas commercial cultivars averaged 18.4 percent. The high groat-protein percentages found for these oat lines is doubly encouraging because (a) some experimental lines from hybridizations among lines of *A. sativa* had protein percentages well above current commercial cultivars and (b) these high-protein experimental lines were extracted from hybrid-derived populations where no consideration was given to crossing parents with genes for high protein percentage.

With rice, sizable increases in grain protein percentage have been made via mutation breeding, i.e., artificially inducing mutations via mutagen treatment. Tanaka and Takagi (1970) reported increasing the protein in rice grain from 6.5 percent in Norin 8 cultivar to 16.0 percent in a mutagen-derived line (Table 22). Note that most of the 500-plus mutagen-derived rice lines were above Norin 8 in protein content.

Taking a cue from the rice studies, we are investigating the protein percentages of mutagen-derived lines of oats. We used 11 oat cultivars and tested equal numbers of mutagen-derived and checklines in each. For each cultivar, the mutagen-derived population contained lines with protein percentages higher than the best checklines, so I will give data from only two cultivars to illustrate the variation induced. For Clintland cultivar, three mutagen-derived lines had

TABLE 22 Frequency Distribution for Grain Protein Percentages of Mutagen-Derived Rice Lines[a]

Protein Percent Class Centers	Number of Lines
4	7
6	165[b]
8	292
10	73
12	12
14	1
16	1

[a]From Tanaka and Takagi (1970).
[b]Parent cultivar, Norin 8, had 6.5 percent grain protein.

TABLE 23 Frequency Distributions and Means of Groat-Protein Percentages for Oat Lines in Mutagen-Derived and Check Populations from Clintland Cultivar[a]

Protein Percentage Class Center	Number of Lines	
	Mutagen-Derived	Check
17	–	6
18	19	22
19	46	43
20	16	13
21	2	1
22	1	–
23	–	–
24	1	–
MEAN	19.2	18.8
CORRELATION[b]	0.00	-0.09

[a]Unpublished data.
[b]Correlations between groat-protein percentages and grain yields.

groat-protein percentages that exceeded the best checkline (Table 23), and no mutagen-derived line was lower than the lowest checkline. The three superior lines, No. 27, 28, 49, from the mutagen-derived population had 21.1, 23.9, and 22.1 percent groat protein, respectively. Line No. 49 had a low grain yield, but the other two yielded better than the check mean. For Burnett cultivar, one mutagen-derived line had a lower and three had higher groat-protein percentages than the lowest and highest checklines, respectively (Table 24). The superior lines, No. 20, 22, and 60 from the mutagen-derived population, had 19.9, 19.2, and 19.4 percent groat protein, respectively. Two lines, No. 22 and 60, produced inferior grain yields, but No. 20 yielded as well as the check mean.

To date, no amino acid analyses have been made on the mutagen-derived lines with high groat protein, but this research is planned. Our studies suggest that grain protein content of oats can be elevated significantly via mutagen treatment.

To transfer high groat-protein genes from *A. sterilis* to commercially acceptable cultivars would appear to be easy since this wild species and cultivated types can be readily crossed. Briggle (1971), however, has cautioned that oat lines he selected from *A. sterilis* X *A. sativa* crosses for high content of groat protein were similar to the wild parent in seed traits, i.e., high hull percentage, awned, pubescent, and slender. Campbell and Frey (1972a), in contrast, found a genetic correlation of 0.07 between groat weight and groat-protein content for 10 *A. sterilis* X *A. sativa* crosses, and they were able to select a sizable number of F_2-derived lines that had high groat-protein and *sativa* seed traits.

Campbell and Frey (1972b) studied the groat-protein percentage among F_2-derived lines of *A. sterilis* X *A. sativa* and found that the inheritance seemed to be relatively simple, i.e., few loci or blocks of loci seemed to account for the segregation patterns. In several crosses, the parental protein percentages were recovered with as few as 35–45 lines tested within crosses.

The data available on grain protein for oats suggest (a) that oat protein has a high biological value relative to other feed grains, (b) that the amino acid composition of groat protein is highly constant for all levels of protein percentage, and (c) that there appear to be several sources of genes for increasing groat-protein percentage. Some current research is devoted to searching for variable amino acid composition in oat protein and to animal feeding tests with grain from different cultivars. The major effort, however, is to simply increase the groat-protein content. Most likely, commercially acceptable oat cultivars with 22–24 percent protein will be available before the end of this decade.

TABLE 24 Frequency Distributions and Means of Groat-Protein Percentages for Oat Lines in Mutagen-Derived and Check Populations of Burnett Cultivar[a]

Protein Percentage Class Center	Number of Lines	
	Mutagen-Derived	Check
12	1	–
13	–	1
14	2	3
15	6	13
16	28	26
17	37	33
18	17	22
19	6	2
20	1	–
MEAN	16.8	16.6
CORRELATION[b]	-0.03	-0.20

[a]Unpublished data.
[b]Correlation between groat-protein percentages and grain yields.

TRITICALE

A new man-made cereal called *Triticale* has received much publicity in the popular press during the past 5–7 years. It resulted from crossing durum wheat (*Triticum durum*) and rye (*Secale cereale*), with subsequent doubling of the hybrid chromosome number to $2n = 42$. *Triticale* and bread wheat (*Triticum aestivum*) are both hexaploid and both carry the *A* and *B* genomes found in durum wheat, but *Triticale* carries the *R* genome from rye whereas bread wheat carries the *D* genome from goat grass (*Aegilops squarrosa*). Replacement of the *D* by the *R* genome has reduced the baking quality of *Triticale* (Unrau and Jenkins, 1964).

The hybrid was made a century ago, but only recently have plant breeders started to exploit its potential for agricultural production. Since rye tends to surpass wheat for protein content in the grain and lysine content in the protein (Block and Weiss, 1956), it has been reasoned that these traits might be transmitted to the hybrid, *Triticale.* Also, Jones *et al.* (1948) have shown the protein of rye has better nutritional quality than does wheat protein (Table 25). Parenthetically, rye is better adapted to marginal production areas, and plant breeders also hope to capture this trait in *Triticale.*

Fox and De Fontaine (1956) reported that both protein and lysine contents of *Triticale* were intermediate between the rye and wheat parents from which the *Triticale* was formed. The lysine content, expressed as percent of protein, was 2.9 and 4.2 in wheat and rye, respectively, and 3.4 for *Triticale.*

Larter (1968) compared the mean composition of six *Triticale* lines with wheat, oats, and barley, all grown in Canada (Table 26). When grain contained 12–13 percent protein, the *Triticale* had a

TABLE 25 Gains in Body Weight, Total and per Gram of Protein, for Rats Fed Wheat and Rye at Several Dietary Protein Levels[a]

Dietary Level of Protein (%)	Total Gain		Gain/Gram Protein	
	Wheat	Rye	Wheat	Rye
4.5	20	35	1.7	2.3
7.5	28	60	1.2	2.2
9.5	66	74	1.6	1.8
MEAN	–	–	1.5	2.1

[a]From Jones *et al.* (1948).

TABLE 26 Protein, Lysine, Methionine, and Threonine Composition of Wheat, Barley, Oat, and *Triticale* Grain[a]

Crop and Cultivar	Protein[b]	Lysine[c]	Methonine[c]	Threonine[c]
Manitou wheat	12.2	2.56	1.77	2.41
Conquest barley	12.1	2.93	1.80	3.10
Harmon oats	10.5	3.52	1.11	3.48
Triticale[d]	12.9	3.20	1.44	3.77

[a]From Larter (1968).
[b]Percent of dry grain.
[c]Percent of protein.
[d]Mean of six lines.

25 percent higher lysine level than did wheat. Methionine was 20 percent lower and threonine was 50 percent higher in *Triticale* than in wheat.

The most extensive survey of *Triticale* grain for protein and lysine composition has been reported by Villegas *et al.* (1968), who analyzed 70 *Triticale* selections and lines and compared them with 25 durum wheat lines (Table 27). Durum wheats averaged 15.8 percent protein in the grain, whereas *Triticale* averaged 17.5 percent. Mean percentage of lysine in the protein was one-tenth higher (2.8 percent for wheat and 3.2 percent for *Triticale*) in *Triticale* than in wheat grain. Not only was lysine percentage in the protein of *Triticale* greater than in wheat, but of more significance, this phenomenon occurs at a higher protein percentage. However, E. M. Larter (personal communication) has shown that whether *Triticale* has a higher percentage of lysine in protein depends upon the strains being analyzed (Table 28).

TABLE 27 Means and Ranges of Protein and Lysine Percentages in Grain of Durum Wheat and Hexaploid *Triticale*[a]

Species	No. Samples	Protein Percentage[b]		Lysine Percentage[c]	
		Mean	Range	Mean	Range
Durum wheat	25	15.8	11.2–20.6	2.8	2.3–3.5
Triticale	70	17.5	12.8–22.5	3.2	2.6–3.7

[a]From Villegas *et al.* (1968).
[b]Expressed as percent of dry weight.
[c]Expressed as percent of protein.

TABLE 28 Conents of Protein, Lysine, Methionine, and Threonine in Grain of Selected Wheat, Rye, and *Triticale* Lines[a]

Species	Name or Number	Percentage of[b]			
		Protein	Lysine	Threonine	Methonine
Wheat	Manitou	17.2	2.59	3.12	0.87
	Pitic	14.6	2.66	2.13	0.91
Rye	White	17.8	2.80	2.09	0.79
Triticale	Rosner	14.4	3.17	2.20	0.84
	Tcl 53	17.9	2.45	2.03	0.77
	Tcl 22	16.7	2.91	2.20	0.82

[a]From E. M. Larter, University of Manitoba, Winnipeg, Manitoba, Canada (unpublished data).
[b]Protein as percent of dry weight and lysine, threonine, and methionine as percents of protein.

When Sell *et al.* (1962) tested *Triticale* as a protein source in rations for 4-week-old chicks, it was similar to barley and wheat in PER (Table 29). However, supplementation of the *Triticale* ration with lysine improved the efficiency of gain materially.

The protein and amino acid composition of *Triticale* grain indicate that this man-made species has promise as a livestock feed, but it does not appear to represent a major breakthrough in improving either protein quantity or quality of cereals. Obviously, plant breeders can select strains with high-protein content, and the lysine level in *Triticale* grain protein averages better than in wheat grain, but lysine supplementation is necessary for optimum livestock gains. Another problem in feeding *Triticale* is its susceptibility to the ergot

TABLE 29 Grams of Feed per Gram of Body-Weight Gain for 4-Week-Old Chicks Fed Rations Composed with Different Cereals as Protein Sources[a]

Grain Source	Percent in Ration	Gram Feed/ Gram Gain
Wheat	67	2.18[b]
Triticale	67	2.14
Barley	60	2.50[b]
Wheat	67	2.38
Triticale	81	2.28

[a]From Sell *et al.* (1962).
[b]Values compared are not significantly different.

fungus (*Claviceps purpurea*). As little as 0.1–0.5 percent ergot sclerotia in *Triticale* can make livestock ill and go off feed.

ASSOCIATION BETWEEN GRAIN YIELDS AND PROTEIN PERCENTAGES

Traditionally, the cereal grains have been grown and utilized as efficient producers of concentrated sources of energy feeds, and the literature abundantly shows that, almost universally, grain yields and protein contents in the grains are negatively correlated (Table 30). The papers listed in Table 30 represent only a meager sample of the literature. Such data prompted Wilcox (1949) to postulate that there exists a "universal nitrogen constant" of 318. By this he means that the maximum amount of nitrogen any species can absorb in a single growth cycle, if all conditions are optimum, is 318 lb/acre. Presumably, this ceiling on nitrogen absorption would lead to negative correlations between yield and nitrogen content even at suboptimal conditions.

Black and Kempthorne (1954) later showed the derivation of the universal nitrogen constant to be in error, and White and Black (1954) showed that, in pot cultures, plants on occasion would absorb much more nitrogen than the equivalent of 318 lb/acre. This led the latter authors to suggest that the seemingly universal negative correlation between grain yields and grain protein percentages was

TABLE 30 Correlations between Grain Yields and Protein Percentages in the Grain Reported by Several Researchers

Crop	Correlation	Reference
Barley	-0.79**	Grant and McCalla (1949)
	-0.24*	Zubriski *et al.* (1970)
Corn	-0.48**	Frey (1951a)
	-0.33*	Dudley *et al.* (1971)
Oats	-0.45**	Jenkins (1969)
Sorghum	-0.85**	Worker and Ruckman (1968)
	-0.26*	Malm (1968)
Wheat	-0.56**	Waldon (1933)
	-0.80**	Grant and McCalla (1949)
	-0.25**	Stuber *et al.* (1962)

**Significant difference at 1 percent level.
*Significant difference at 5 percent level.

TABLE 31 Grain Yields and Protein Percentages of Wheat Cultivars Tested for 3 Years in Southeastern Wheat Nursery[a]

Cultivar	Grain Yield (q/ha)	Protein (%)
Hardired 47-12	1.8	10.1
Chancellor	1.8	10.9
Purcam	1.8	11.4
Coker 47-27	1.9	11.6
Taylor	1.9	11.9
Atlas 50	1.9	12.8
Atlas 66	1.9	13.3

[a]From Middleton *et al.* (1954).

an artifact of inadequate availability of soil nitrogen in most experiments. The implication is that the reported negative correlations are phenotypically real, but not genotypic in origin.

Although most literature is to the contrary, Middleton *et al.* (1954) reported two wheat cultivars, Atlas 50 and Atlas 66, that in extensive tests (3 years and many sites) were equally as productive as older cultivars, and yet they had from 0.9 to 3.2 percent more protein in their grain (Table 31). The genes that gave this increase in protein content were from Frondosa, a Brazilian cultivar. Johnson *et al.* (1971), using Atlas 66 as a source of genes for high protein, have isolated several "second cycle" wheat lines that are hard red winter types and combine high grain yields with a 2.5 percent increase in protein content (Table 32).

At the Iowa station, we have tested oat cultivars and lines in field experiments where soil nitrogen was very deficient and very adequate,

TABLE 32 Grain Yields and Protein Percentages of Wheat Lines Tested at Three Sites in Nebraska in 1970[a]

Wheat Line	Grain Yield (q/ha)	Protein (%)
Scout 66 (check)	3.4	11.8
NB 701132	3.9	14.4
NB 701134	3.6	14.3
NB 701137	3.6	14.1
NB 701154	3.5	14.4

[a]From Johnson *et al.* (1971a).

TABLE 33 Ranges of Grain Yields and Protein Percentages and Correlations between These Two Traits for 60 Oat Lines and Cultivars When Grown under Conditions of Deficient and Adequate Soil Nitrogen[a]

Soil-Nitrogen Condition	Range of Protein Percentages	Range of Grain Yields	Correlation
Deficient	15.0–20.0	10–21	–0.26*
Adequate	16.0–20.6	11–42	+0.04

[a] Unpublished data.
*Significant difference at 5 percent level.

respectively, for growth of oat plants. At the latter site, 80 kg/ha of available nitrogen remained after the crop was mature. In these experiments, the correlations between grain yields and grain protein percentages were –0.24* and + 0.04, respectively (Table 33).

Seemingly, the more-or-less universal reports of a negative correlation between grain yields and protein percentages for cereal grain crops is an artifact of past experimentation. In wheat at least, some new cultivars have yields equivalent to the protein percentages one-fifth greater than the old cultivars. The data from oats indicate that nitrogen fertilization commonly used for small grains may need to be increased if the genetic potentials for high yields and high-protein percentage in new cultivars are to be exploited.

SUMMARY

Cereal crop researchers in the past, whether breeding new cultivars or formulating new crop husbandries, have placed most emphasis on increasing total grain yields. Chemical compositions of the seeds of cereal grains used for livestock feed have been treated academically (e.g., several hundred scientific papers have been published on the starches and proteins of corn grain alone), but little of this information has been used systematically in attempts to improve the quality of cereal grains as livestock feed. And there are some who say that this is as it should be, i.e., cereal grain crops are efficient producers of sources of concentrated energy and this trait should be exploited to its utmost. These persons would argue that the protein needs in livestock rations can be met more easily from other crop plants that are efficient protein producers.

Time alone will tell whether it is better to breed separate crop plants to meet energy and protein needs, or to breed cereal crops to meet both needs. However, as is evident from this review, extensive research money and effort currently are being expended to improve the quantity and/or quality of proteins in the seeds of cereal crops. All of the cereal grains, except oats, show a negative relationship between quantity of protein in the grain and the biological quality of the protein. With barley and corn, mutant endosperm types have been found that change the composition among essential amino acids of the proteins of these two grains from lysine-deficient to lysine-adequate for monogastric animals. And, if Vavilov's "principle of homologous series" holds true, it is likely that similar mutant forms will be found within the world collections of other cereal species, i.e., sorghum and wheat. For wheat, commercial cultivars with increased protein content are already available. For livestock feed, it may be that a new man-made species, *Triticale,* will provide a somewhat better quality of protein than that in wheat, by including genes for high lysine from rye. Oats are unusual in that their protein is of relatively good biological value, and the protein quality is constant at all levels of protein in the grain; therefore, emphasis with oats is simply to increase the grain protein content. The seemingly universal negative correlation between protein content of grain and grain yield per hectare is probably an artifact, resulting from inadequate nitrogen availability for growing plants.

It is likely that new cultivars of wheat and oats with higher protein content and new cultivars of corn and barley with higher quality protein will be available for commercial use before the end of this decade.

REFERENCES

Alexander, D. E., R. J. Lambert, and J. W. Dudley. 1969. Breeding problems and potential of modified protein in maize, p. 55–65. *In* new approaches to breeding for improved plant protein. STI/PUB/212. International Atomic Energy Agency, Vienna, Austria.

Baumgarten, W., A. N. Mather, and L. Stone. 1946. Essential amino acid composition of feed materials. Cereal Chem. 23:135–155.

Beeson, W. M., R. A. Pickett, E. T. Mertz, G. L. Cromwell, and O. E. Nelson. 1966. Nutritive value of high-lysine corn, p. 70. *In* Proceedings, Distillers feed research conference, Cincinnati, Ohio.

Bishop, L. R. 1928. The composition and quantitative estimation of barley proteins. J. Inst. Brewing 34:101.

Black, C. A., and O. Kempthorne. 1954. Wilcox's agrobiology: II. Application of the nitrogen constant 318. Agron. J. 46:307–310.

Block, J. R., and K. W. Weiss. 1956. Amino acid handbook. Methods and results of protein analyses. Charles C Thomas, Springfield, Illinois.

Briggle, L. W. 1971. Improving nutritional quality of oats through breeding. Agron. Abstr. 1971. p. 53.

Campbell, A. R., and K. J. Frey. 1972a. Associations between groat-protein percentage and certain plant and seed traits in interspecific oat crosses. Euphytica 21:353–362.

Campbell, A. R., and K. J. Frey. 1972b. Inheritance of groat-protein in interspecific oat crosses. Can. J. Plant Sci. 52:735–742.

Collins, F. C., and R. C. Pickett. 1972. Combining ability for yield, protein, and lysine in an incomplete diallel of *Sorghum bicolor* (L.) Moench. Crop Sci. 12:5–6.

Cromwell, G. L., R. A. Pickett, and W. M. Beeson. 1967. Nutritive value of *opaque-2* corn for swine. J. Anim. Sci. 26:1325.

Cromwell, G. L., R. A. Pickett, T. R. Cline, and W. M. Beeson. 1969. Nitrogen balance and growth studies of pigs fed *opaque-2* and normal corn. J. Anim. Sci. 28:478–483.

Csonka, F. A. 1937. Amino acids of staple foods. I. Wheat (*Triticum vulgare*). J. Biol. Chem. 118:147–153.

Deosthale, Y. G., V. S. Mohan, and K. V. Rao. 1970. Varietal differences in protein, lysine, and leucine content of grain sorghum. J. Agric. Food Chem. 18:644–646.

Deyoe, C. W., and J. A. Shellenberger. 1965. Amino acids and proteins of sorghum grain. J. Agric. Food Chem. 13:446–450.

Dudley, J. W., R. J. Lambert, and D. E. Alexander. 1971. Variability and relationships among characters in *Zea mays* L. synthetics and improved protein quality. Crop Sci. 11:512–514.

Emerson, R. A., C. W. Beadle, and A. C. Fraser. 1935. A summary of linkage studies in maize. Cornell Univ. Agric. Sta. Mem. 180.

Ewart, J. A. D. 1967. Amino acid analyses of cereal flour proteins. J. Sci. Food Agric. 18:548–552.

Feist, H. A., and R. J. Lambert. 1970. Changes in six different *opaque-2* genotypes of *Zea mays* during successive generations of backcrossing. Crop Sci. 10:663–665.

Folkes, B. N., and E. W. Yemm. 1956. Amino acid content of the proteins of barley grain. Biochem. J. 62:4–11.

Fonseca, J. B., J. C. Rogler, W. R. Featherston, and T. R. Cline. 1969. Nutritional evaluation of *opaque-2* corn and safflower meal in poultry rations. Poult. Sci. 48:1807.

Fonseca, J. B., J. C. Rogler, W. R. Featherston, and T. R. Cline. 1970. Further studies on the nutritive value of *opaque-2* corn for the chick. Poult. Sci. 49:1518–1525.

Fox, S. W., and D. De Fontaine. 1956. Sequential assays of four amino acids in proteins of wheat, rye, and their hybrids. Proc. Soc. Exp. Biol. Med. 92:503–506.

Frey, K. J. 1948. Inheritence of protein, zein, tryptophan, valine, leucine, and isoleucine in two maize hybrids. Ph.D. thesis. Iowa State University, Ames.

Frey, K. J. 1951a. The interrelationships of protein and amino acids in corn. Cereal Chem. 28:123–132.

Frey, K. J. 1951b. The relation between alcohol-soluble and total nitrogen content in oats. Cereal Chem. 28:506–509.

Frey, K. J., and G. I. Watson. 1950. Chemical studies on oats. I. Thiamine, niacin, riboflavin, and pantothenic acid. Agron. J. 42:434–436.

Frey, K. J., B. Brimhall, and G. F. Sprague. 1949. Effects of selection on protein quality in the corn kernel. Agron. J. 41:399–403.

Fuchs, J. A. 1970. Genetic variation in lysine and protein content in corn seed. Ph.D. thesis. Texas A&M University, College Station.

Grant, M. N., and A. G. McCalla. 1949. Yield and protein content of wheat and barley. I. Interrelations of yield and protein content of random selections from single crosses. Can. J. Res. (Section C) 27:230–240.

Hagberg, A., and K. E. Karlsson. 1969. Breeding for high-protein content and quality in barley, p. 17–21. *In* New approaches to breeding for improved plant protein. STI/PUB/212. International Atomic Energy Agency, Vienna, Austria.

Haikerwal, M., and A. R. Mathieson. 1971. The protein content and amino acid composition of sorghum grain. Cereal Chem. 48:690–699.

Hansen, D. W., B. Brimhall, and G. F. Sprague. 1946. Relationship of zein to total protein in corn. Cereal Chem. 23:329–335.

Harpstead, D. D. 1971. High-lysine corn. Sci. Am. 225:34–42.

Hegsted, D. M., M. F. Trulson, and F. J. Stare. 1954. Role of wheat and wheat products in human nutrition. Phys. Rev. 34:221–258.

Hischke, H. H., Jr., G. C. Potter, and W. R. Graham, Jr. 1968. Nutritional value of oat protein. I. Varietal differences as measured by amino acid analyses and rat growth responses. Cereal Chem. 45:374–378.

Hopkins, C. G. 1899. Improvement of chemical composition of the corn kernel. Ill. Exp. Sta. Bull. 55. p. 205–240.

Howe, E. E., G. R. Jansen, and E. W. Gilfillan. 1965. Amino acid supplementation of cereal grains as related to the world food supply. Am. J. Clin. Nutr. 16:315–320.

Jenkins, G. 1969. Grain quality and hybrids of *Avena sativa* L. and *A. byzantina* C. Koch. J. Agric. Sci. (Cambridge) 72:311–317.

Johnson, V. A., J. W. Schmidt, P. J. Mattern, and A. Haunold. 1963. Agronomic and quality characteristics of high-protein F_2-derived lines from a soft red winter–hard red winter wheat cross. Crop Sci. 3:7–10.

Johnson, V. A., P. J. Mattern, and J. W. Schmidt. 1967. Nitrogen relations during spring growth in varieties of *Triticum aestivum* L. differing in grain protein content. Crop Sci. 7:664–667.

Johnson, V. A., P. J. Mattern, D. A. Whited, and J. W. Schmidt. 1969. Breeding for high protein content and quality in wheat, p. 29–40. *In* New approaches of breeding for improved plant protein. STI/PUB/212. International Atomic Energy Agency, Vienna, Austria

Johnson, V. A., P. J. Mattern, and J. W. Schmidt. 1971. Genetic studies of wheat protein. Paper presented at American Chemical Society Symposium on seed protein. Los Angeles, California. Mar. 29–31, 1972.

Johnson, V. A., P. J. Mattern, and J. W. Schmidt. 1972a. Genetic studies of wheat protein. *In* Symposium: Seed proteins. G. E. Inglett (ed.), p. 126–135. Avi Publishing Co., Inc. Westport, Connecticut. 313 p.

Johnson V. A., P. J. Mattern, and J. W. Schmidt. 1972b. Wheat protein improvement. *In* Rice breeding, p. 407–408. International Rice Research Institute, Los Banos, Philippines. 738 p.

Jones, D. B., A. Caldwell, and K. D. Widness. 1948. Comparative growth-promoting values of the proteins of cereal grains. J. Nutr. 35:639–649.

Lambert, R. J., D. E. Alexander, and J. W. Dudley. 1969. Relative performance of normal and modified protein *opaque-2* hybrids. Crop Sci. 9:242–243.

Larter, E. M. 1968. *Triticale.* Agric. Inst. Rev. (Ottawa) 23:12–15.

Lawrence, J. M., K. M. Day, E. Huey, and B. Lee. 1958. Lysine content of wheat varieties, species, and related genera. Cereal Chem. 35:169–178.

Leng, E. R. 1961. Predicted and actual responses during long-term selection for chemical composition in maize. Euphytica 10:368–378.

Malm, M. R. 1968. Exotic germplasm use in grain sorghum improvement. Crop Sci. 8:295–298.

Marias, J. S. C., and D. B. Smuts. 1940. The biological value of proteins of maize and maize supplemented with lysine and tryptophan. Onderstepoort J. Vet. Med. 15:197–205.

Mattern, P. J., A. Salem, V. A. Johnson, and J. W. Schmidt. 1968. Amino acid composition of selected high-protein wheats. Cereal Chem. 45:437–444.

McWhirter, K. S. 1971. A floury endosperm high-lysine locus on chromosome 10. Maize Gen. Coop. Newsl. No. 45. Bloomington, Indiana, p. 184.

Mertz, E. T., L. S. Bates, and O. E. Nelson. 1964. Mutant gene that changes protein composition and increases lysine content of maize endosperm. Science 145:279–280.

Mertz, E. T., O. E. Nelson, L. S. Bates, and O. A. Veron. 1966. Better protein quality in maize. *In* R. F. Gould [ed.] World protein sources. Adv. Chem. Ser. 56:228–242.

Mertz, E. T., O. A. Veron, L. S. Bates, and O. E. Nelson. 1965. Growth of rats fed on *opaque-2* maize. Science 148:1741–1742.

Middleton, G. K., C. E. Bode, and B. B. Bayles. 1954. A comparison of the quantity and quality of protein in certain varieties of soft wheat. Agron. J. 46:500–502.

Millder, D. F. 1958. Composition of cereal grains and forages. Publ. 585. National Academy of Sciences–National Research Council, Washington, D.C. 658 p.

Mitchell, H. H., and R. J. Block. 1946. Some relationships between amino acid contents of proteins and their nutritive values for the rat. J. Biol. Chem. 163:599–620.

Morkuze, Z. 1937. Biological value of the proteins of certain cereals. Biochem. J. 31:1973–1977.

Mossé, J. 1966. Alcohol-soluble proteins of cereal grains. Fed. Proc. 25:1663–1669.

Munck, L. 1971. High lysine barley—A summary of the present research development in Sweden. Barley Gen. Newsl. 2:54–59.

Munck, L., A. E. Karlsson, and A. Hagberg. 1969. Selection and characterization of a high-protein, high-lysine variety from the World Barley Collection, p. 544–558. *In* Proceedings of the second international barley general symposium. Washington State University Press, Pullman, Washington.

Murphy, J. H., and A. Dalby. 1971. Changes in the protein fractions of developing normal and *opaque-2* maize endosperm. Cereal Chem. 48:336–349.

Nelson, O. E. 1969. The modification by mutation of protein quality in maize, p. 41–54. *In* New approaches to breeding for improved plant protein. STI/PUB/212. International Atomic Energy Agency, Vienna, Austria.

Nelson, O. E., E. T. Mertz, and L. S. Bates. 1965. Second mutant gene affecting the amino acid pattern of maize endosperm protein. Science 150:1469–1470.

Osborne, T., and S. H. Clapp. 1908. Hydrolysis of proteins of maize. Am. J. Phys. 20:447–493.

Osborne, T., and L. B. Mendel. 1914. Nutritive properties of proteins of the maize kernel. J. Biol. Chem. 18:1–16.

Pond, W. G., J. C. Hillier, and G. A. Benton. 1958. The amino acid adequacy of milo (grain sorghum) for the growth of rats. J. Nutr. 65:493–502.

Robbins, G. S., Y. Pomeranz, and L. W. Briggle. 1971. Amino acid composition of oat groats. Agric. Food Chem. 19:536–539.

Showalter, M. F., and R. H. Carr. 1922. The characteristic proteins of high- and low-protein corn. J. Am. Chem. Soc. 44:2019–2023.

Sell, J. L., G. C. Hodgson, and L. H. Shebeski. 1962. *Triticale* as a potential component of chick rations. Can. J. Anim. Sci. 42:158–166.

Sreeramulu, C., and L. F. Bauman. 1970. Yield components and protein quality of *opaque-2* and normal diallels in maize. Crop Sci. 10:262–265.

Sreeramulu, C., L. F. Bauman, and G. Roth. 1970. Effect of outcrossing on protein quality, kernel weight, and related characters in *opaque-2* and *floury-2* maize (*Zea mays*). Crop Sci. 10:235–236.

Stephenson, E. L., J. O. York, D. B. Bragg, and C. A. Ivy. 1971. The amino acid content and availability of different strains of grain sorghum to the chick. Poult. Sci. 50:581–584.

Stuber, C. W., V. A. Johnson, and J. W. Schmidt. 1962. Grain protein content and its relationship to other plant and seed characters in the parents and progeny of a cross of *Triticum aestvium.* Crop Sci. 2:502–508.

Tanaka, S., and Y. Takagi. 1970. Protein content of rice mutants, p. 55–62. *In* Improving plant protein by nuclear techniques. SM132/2. International Atomic Energy Agency, Vienna, Austria.

Unrau, A. M., and B. C. Jenkins. 1964. Investigations of synthetic cereal species. Milling, baking, and some compositional characteristics of some *Triticale* and parental species. Cereal Chem. 41:365.

Vavich, M. G., A. R. Kemmerer, B. Nimbkar, and L. S. Stith. 1959. Nutritive value of low- and high-protein sorghum grains for growing chickens. Poult. Sci. 38:36–40.

Villegas, E., C. E. McDonald, and K. A. Gilles. 1968. Variability in the lysine content of wheat, rye, and *Triticale* proteins. Centro de Mejoramiento de Maiz ej Trigo Res. Bull. No. 10. 31 p.

Villegas, E., C. E. McDonald, and K. A. Gilles. 1970. Variability of the lysine content of wheat, rye, and *Triticale* proteins. Cereal Chem. 47:746–757.

Viuf, B. T. 1969. Breeding of barley varieties with high-protein content with respect to quality, p. 23–28. *In* New approaches to breeding for improved plant protein. STI/PUB/212. International Atomic Energy Agency, Vienna, Austria.

Virupaksha, T. K., and L. V. S. Sastry. 1968. Studies on the protein content and the amino acid composition of some varieties of grain sorghum. J. Agric. Food. Chem. 16:199–203.

Waggle, D., D. B. Parrih, and C. W. Deyoe. 1966. Nutritive value of protein and high and low protein content sorghum grain as measured by rat performance and amino acid assay. J. Nutr. 88:370–374.

Waldon, L. R. 1933. Yield and protein content of hard red spring wheat under conditions of high temperature and low moisture. J. Agric. Res. 47:129–149.

Weber, E. A., J. P. Thomas, R. Reder, A. M. Schlehuber, and D. A. Benton. 1957. Protein quality of oat varieties. J. Agric. Food Chem. 5:926–928.

White, W. C., and C. A. Black. 1954. Wilcox's agrobiology: III. The inverse yield-nitrogen ratio. Agron. J. 46:310–313.

Wilcox, O. W. 1949. Keys to abundance. Better Crops Plant Food 33 (4):9–15, 46–48.

Willcock, E. G., and F. G. Hopkins. 1913. Importance of amino acids in metabolism. J. Phys. 35:88–101.

Worker, G. F., Jr., and J. Ruckman. 1968. Variations in protein levels in grain sorghum grown in the Southwest desert. Agron. J. 60:485–488.

Zubriski, J. C., E. H. Basey, and E. D. Norum. 1970. Influence of nitrogen and potassium fertilizers and dates of seeding on yield and quality of malting barley. Agron. J. 62:216–219.

G. O. Kohler, Joseph Chrisman, and E. M. Bickoff

SEPARATION OF PROTEIN FROM FIBER IN FORAGE CROPS

Green leaf crops are the major source of protein for livestock in this country. The total hay crop, without considering forage consumed in pastures and range or conserved as silage, amounted, in 1970, to about 128 million tons. Of this tremendous hay crop, about three fifths, or 75 million tons, is alfalfa. This alfalfa hay crop contains as much protein (11.2 million tons) as the entire 1.14 billion bushel soybean crop or the 4.1 billion bushel corn crop.

It has been a research goal in many laboratories to develop improved products from forages designed for specific types and classes of animals. These processes would be based, as is dehydration, on fresh leafy green plants recovered essentially quantitatively from the field and would seek to eliminate the losses inherent in haying and ensiling. The processes under investigation are aimed at fractionating the forage into (a) high-protein products for special purpose feed uses or for human food uses and (b) a product for ruminants equivalent in quality to high quality hay or silage that could compete with these products in price. The new ruminant products, like dehydrated alfalfa, could be made to order and would have guaranteed composition and specifications.

In this discussion, I shall be referring to alfalfa almost continuously because most of the research on processing to higher protein, special purpose products are based upon it. The rationale for this approach is as follows:

- Alfalfa is the major forage crop, hence many varieties are available to fit differing conditions and agronomic practices.
- It survives harvesting at frequent intervals.
- It maintains relatively high nutritional quality throughout hot and cool seasons, so long as adequate moisture is available.
- It is a perennial crop, hence time and costs of replanting are minimal.
- It is a legume that, with the aid of symbiotic nodule bacteria, fixes nitrogen and fits well in crop rotation systems.
- There is already a substantial processing industry that has solved many problems relating to farmers, i.e., harvesting and hauling, etc.
- It produces more protein per acre than any other crop.

The data in Table 1 illustrate this last point by comparing dry matter and protein produced per acre in three locations with the average values for soybeans. It will be noted that alfalfa leads soybeans in all cases so far as crude protein is concerned. The areas selected are ones where alfalfa is currently grown and dehydrated.

Having shown why alfalfa is selected as the raw material of choice, let us pose the questions, What are the weaknesses of alfalfa and other forage crops as feeds that justify research on upgrading them? What possible economic advantage could be gained by making high- and low-fiber products from alfalfa?

Let us first consider the basic instability of the fresh plant and the losses that occur during haying and ensiling. The fresh alfalfa plant

TABLE 1 Alfalfa as a Raw Material for Leaf Protein Concentrate

Geographical Locality	No. of Cuttings/ Season	Annual Yield (tons/acre)	Crude[a] Protein (%)	Crude Protein (lb/acre)
Nebraska	4	4	18.5	1,500
California				
Southern Imperial Valley	10	10	20	4,000
Northern Central Valley	7	7	20	2,800
		1.76[b]	25[b]	1,433[b]

[a] 93 percent solids.
[b] Comparable values for soybeans.

growing in the field usually contains 70–80 percent moisture. On cutting, it rapidly undergoes enzymatic and microbial changes that reduce or destroy its value as a feed. Sun drying (haying) has been used since antiquity to conserve green crops, but the process is wasteful. Losses may range from 15 to over 40 percent of the dry matter of the crop, and these losses are greatest for the most nutritious parts of the plants—the leaves. Ensiling, which was described as early as the eighteenth century, is also wasteful since much of the carbohydrate and protein may be lost during fermentation in the silo, by surface spoilage and by liquid drainage (Shepherd *et al.*, 1954). The most effective process yet developed for converting the fresh plant nutrients into a stable form is dehydration. Here over 95 percent of the dry matter of the fresh plant is recovered in the form of a high quality product. But, even though more of the crop is recovered, the costs involved in carrying out the dehydration operation have not permitted the product so derived to replace hay and silage as primary roughage and energy sources for most animals. Hence, dehydrated alfalfa is used in special-purpose feeds as (a) a source of xanthophyll for poultry, (b) a source of unidentified growth and reproduction factors (UGF) for poultry, swine, and ruminants, (c) a source of unidentified urea utilization factor (UUF) for ruminants, and (d) range supplements for ruminants.

The economic rationale for fractionation procedure lies in the fact that alfalfa, at 15–22 percent protein and 18–35 percent fiber, lies intermediate between a roughage and a concentrate. It is extremely variable. As a roughage in ruminant rations, alfalfa products must compete with such cheap roughages as sawdust, oilseed hulls, and sugar cane bagasse. As an energy source it must compete with corn, milo, and blackstrap molasses, which are cheaper than alfalfa hay per unit of energy. As a nitrogen source, it must compete with cheap protein sources or, in the case of ruminants, with cheap nonprotein nitrogen (NPN) sources. If the feed formulator wants to meet the animal's requirements at minimal cost, he needs accurate knowledge of the composition and specifications of his ingredients, the prices of all available ingredients, and certain limiting levels. His computer then tells him what to do. Where the energy, roughage factor, and protein are in separate packages, the computer has more degrees of freedom to meet the program restrictions than when these nutrients are in a single product such as whole alfalfa. By fractionation, we can see the potential of making products designed for specific end uses, thus increasing the overall intrinsic value of the crop.

Let us next consider three nutrient distribution features of alfalfa that might provide the technological basis for fractionation to yield high- and low-protein products. These are (a) the leaves are richer in protein and lower in fiber than the stems (Chrisman *et al.*, 1971; Kohler and Chrisman, 1966), (b) the tops of the plants contain more protein (and leaves) than the bottoms (Whitney and Hall, 1963; Ayres, 1968; Klimes and Verosta, 1962; Clarke, 1938; Ogden and Kehr, 1968), and (c) the contents of the cells of both leaves and stems are richer in protein than the cell walls (Nageli, 1866). Each of these has provided the basis for process researches that are next discussed in some detail.

Leaf–Stem Separation in Fresh Alfalfa

Some years ago, work was undertaken by L. F. Whitney at Michigan State University and later by G. E. Ayres at Iowa State University to harvest the leaves by stripping them from the standing plant in the field. The plan included dehydrating the leaves in a conventional drum dryer and leaving the stems in the field to grow a second and possibly a third and fourth crop of leaves for later stripping. An alternative approach was to harvest the stem after stripping leaves for green-chop feeding, or to cut and sun dry them to make stem hay. The first experimental leaf-stripping machine that was made and tested was developed from a commercial hay conditioner. An overhead belt-type draper was mounted ahead of the rolls to bend the alfalfa back so that the tops fed into the rolls first. A rubber mat was fastened around the upper roll of the hay conditioner to strip the leaves off the stems as they were pulled between the rolls.

A larger machine was made by mounting the rolls and draper on the front of flail-type forage harvester. Table 2 shows some data reported by Ayres (1968) in which one-leaf-harvest and two-leaf-harvest systems were compared, using bud stage and one-tenth bloom stage alfalfas. It will be noted that only about half as much leaf was obtained as compared with stem yield.

Table 3 shows the analyses of the fractions obtained (Ayres, 1968). The second leaf crop from the one-tenth bloom alfalfa had a disappointingly low-protein content. While the ideas involved here are highly interesting and show promise, a great deal more effort will be needed to perfect equipment capable of commercial operation. Work on leaf-stripping equipment and logistics is being continued at

TABLE 2 Average Dry Matter and Crude Protein Yields per Acre with Two Harvest Systems[a]

	Bud Stage (lb/acre)		One-Tenth Bloom (lb/acre)	
Material	Dry Matter	Protein	Dry Matter	Protein
One-Leaf-Harvest System				
Stripped leaves	293	82	1,021	228
Stems	713	129	2,144	302
Control	1,102	232	2,998	508
Two-Leaf-Harvest System				
First leaves	318	88	1,056	233
Second leaves	491	123	703	131
Stems	1,376	230	3,109	401
Control	1,102	232	2,998	508

[a]From Ayres (1968).

the University of Missouri by H. D. Currence (personal communication), who is building a new stripping machine.

Leaf–Stem Separation of Dried Alfalfa

Many attempts have been made since the early days of the dehydration industry to obtain high-protein fractions by screening dehydrated ground alfalfa. While some were partially successful and resulted in commercial installation of sifters, most of the latter are no longer in use. Drawbacks included inadequate quality differentials between

TABLE 3 Average Crude Protein and Crude Fiber Percentages with Two Harvest Systems[a]

	Bud Stage (%)		One-Tenth Bloom (%)	
Material	Protein	Fiber	Protein	Fiber
One-Leaf-Harvest System				
Stripped leaves	28.2	12.15	22.2	14.54
Stems	18.1	26.11	14.0	32.20
Control	21.2	20.54	17.0	25.63
Two-Leaf-Harvest System				
First leaves	27.7	11.89	22.1	14.17
Second leaves	25.0	15.13	18.6	18.48
Stems	16.7	28.87	12.0	33.47
Control	21.2	20.54	17.0	25.63

[a]From Ayres (1968).

TABLE 4 Lahontan Alfalfa Plant Composition

Plant Part	Weight (%)	Protein (%)	Fiber (%)
Leaf midribs	5.4	30.3	11.1
Leaf, less midribs	40.6	37.3	7.0
Petioles	5.2	21.2	17.6
Terminal buds	5.4	38.1	13.2
Epidermis	8.9	17.3	24.2
Stem, less epidermis	34.5	13.2	40.7
AVERAGE	–	26.0	21.3

fractions, high maintenance and operation costs due to blinding of sieves, and difficulties in marketing products at a high enough price. Some years ago we undertook to reinvestigate the problem from the standpoint of technology and economics. Initially, we carried out a number of hand dissections to determine the composition of various parts of the plant. Table 4 shows the results of a run made on 10–14-in. tall prebud plants of Lahontan alfalfa cut in mid-June (Chrisman *et al.*, 1971). It will be noted that the raw material contained 26.0 percent protein and 21.3 percent fiber. The leaf parts and buds showed very high-protein and low-fiber values. These together with the petioles would make up the major portion of a mechanically separated leaf fraction amounting to 56.6 percent of the dry weight and containing 35.2 percent protein. In our subsequent work, we have continued hand dissections to provide a basis for estimating the efficiency of separations, reducing the number of fractions separated to two. The principles involved in dry separation of leaf from stem are as follows: The dried alfalfa chops, as they emerge from the dehydrator, show a differential in moisture content between the leaf and the stem. The leaves, being only 10–15 cells thick, dry very rapidly and should emerge from the dryer containing 2–5 percent moisture. The moisture of the stems must pass through a longer diffusion path to escape. Hence, the stems emerge containing two to three times as much moisture as the leaves. As the alfalfa pieces enter the dehydrator, most of the leaves are still attached to the stems and hence more rapid passage of the lighter leaf through the drum is limited. Since friability of both leaf and stem is inversely related to moisture content, leaves subjected to impact at the point of discharge from the dehydrator are broken up, while the wetter, tougher stems are not. After this differential in particle size is attained, separation of leaf

FIGURE 1 Mobile dehydrator, Western Regional Research Laboratory, USDA.

from stem may be carried out either by screening or by air separation.

The establishment of these principles and their application required laboratory, pilot plant, and finally commercial-scale research. The project was supported by the State of Nebraska and carried out with the cooperation of members of the American Dehydrators Association.

Because it seemed important to make the separation immediately after rapid dehydration in order to prevent equilibration of moisture in the product, it was necessary first to develop a pilot plant facility for dehydration. Figure 1 shows our installation on a 50-ft bed trailer. It included a vertical-type Rietz disintegrator with a ¾-in. or 1-in. screen. Initially, a 12-mesh shaker screen was used for the separation. A typical experiment carried out at Dixon, California, using this equipment is summarized in Table 5. Three plots, large enough for harvesting with a commercial-scale harvester, were laid out. The plots were cut throughout an entire season at intervals of 26, 30, and 34 days, respectively. The results show that yield of alfalfa and of protein per acre for the season increased as cutting cycle was lengthened. The tonnage of fine (or leaf-enriched) fraction increased from 2.4 tons/acre at 26 days to 2.95 tons/acre for the 30-day cycle. No further increase occurred in the 34-day-cycle material. The protein

TABLE 5 Lahontan Alfalfa from California, 1964 Crop: Yield, Distribution, and Protein Content of Three Cutting Cycles

	26-Day Cycle Plot 1, 7 Cuts		30-Day Cycle Plot 2, 7 Cuts		34-Day Cycle Plot 3, 6 Cuts	
Alfalfa	Tons/ Acre	Protein (%)	Tons/ Acre	Protein (%)	Tons/ Acre	Protein (%)
Whole alfalfa	5.44	22.1	6.59	21.8	7.41	19.9
Fine	2.14	28.0	2.95	27.3	2.94	26.2
Coarse	3.29	16.9	3.64	16.3	4.47	14.2

percentage in the leaf fraction was over 26 percent even at the 34-day cycle. Very large increases occurred in the coarse (stem) fraction as the time between cuttings increased. The quality of the coarse fraction dropped as cutting cycle increased, but at the 34-day point it was still above 14 percent protein, which is comparable to much of the hay produced and is adequate for growing rations, roughages in high energy rations, etc. (Kohler and Chrisman, 1966).

One approach to fractionation of the coarse and fine particles after differential milling is air separation. Here the leaf-shattered crops are passed into the air separator by means of a distributing conveyor and a rotary valve. Air is drawn up through a zig-zag column and carries with it the smaller particles (leaf) and allows the coarser, heavier particles to drop through the bottom of the column. The air carrying the fine portion is passed through a cyclone separator equipped with a rotary valve at the bottom through which the fine fraction is discharged from the cyclone. Over a 3-year period, experiments were carried out on a laboratory scale, a pilot plant scale, and, finally, in Nebraska, on a commercial scale. The operating characteristics were determined for alfalfa and data were obtained for a cost analysis by cooperating workers of the Economic Research Service. It was found that, of the many factors that affected the degree of separation obtained, the feed rate and the air velocity were dominant. Under conditions not too far from optimal (Table 6), a raw material with equivalent to 16.6 percent protein was separated into 44.3 percent of a leaf-rich fraction containing 23.0 percent protein and 24.8 percent fiber and 55.7 percent of stem-rich fraction with 11.5 percent protein and 41.0 percent fiber.

Hand dissection of the same raw material showed that only 26.9 percent of the dry weight actually was leaf. This and a comparison of the protein and fiber contents shows that under the conditions

TABLE 6 Comparison of Two Methods for Fractionation of Harvested Alfalfa

	Weight (%)	Protein (%)	Fiber (%)	Recovery (%)	
				Protein	Fiber
Dehydrated–Air Separated					
Leaf	44.3	23.0	24.8	61.5	32.5
Stem	55.7	11.5	41.0	38.5	67.5
Whole	100.0	16.6	33.8		
Raw–Hand Separated					
Leaf	26.9	29.9	11.5	49.7	9.8
Stem	73.1	11.2	38.9	50.3	90.2
Whole	100.0	16.2	31.5		

used, considerable amounts of stem were included in the leaf-rich fraction. On the other hand, the stem-rich fraction in this run contained little leaf material. It should be emphasized at this point that by a simple control of an air valve the composition of the fractions can be readily shifted.

Table 7 shows a series of five runs in which maximum differential was sought. The air velocity was 550 linear ft/min. The mean raw material ran 19.2 percent protein and 27.5 percent fiber. There were obtained 31.6 percent leaf fraction (26.4 percent protein, 17.2 percent fiber) and 68.4 percent stem fraction (15.8 percent protein, 33.1 percent fiber). The xanthophyll, an important pigment for poultry, is concentrated in the leaf fraction in about the same proportion as the protein. We were able to pellet the stem fraction without grinding using a ½- in. X 2-in. die or a ¾-in. square die.

Since the technical feasibility of dry separation of leaf from stem seemed established, we enlisted the help of the Economic Research Service to run a cost analysis. While the overall analysis is not yet completed, the following quotations are from the summary of the

TABLE 7 Composition of Dehydrated, Air-Separated Alfalfa Hay[a]

	Weight (%)	Protein (%)	Fiber (%)	Carotene (mg/lb)	Xanthophyll (mg/lb)
Feed	100.0	19.2	27.5	99.6	124.5
Leaf	31.6	26.4	17.2	155.1	165.1
Stem	68.4	15.8	33.1	75.8	105.8
% RECOVERY		44.1	19.8	46.7	41.8

[a]Average of five runs at air velocity of 550 ft/min.

report on production costs for dehydration of alfalfa and the added costs necessary to include the air separation process (Vosloh, 1970).

Costs of the new separation process were synthesized for six model dehydrating plants, using the economic-engineering technique.

Investment costs for the standard dehydration plant models–without separation–ranged from about $190,000 for the smallest plant, producing 5,000 tons a year, to about $320,000 for one producing 17,000 tons. Equipment and facilities for separation increased model costs by $64,000–$81,500. Operating costs ranged from $18.37 a ton for the smallest plant to $11.01 for the largest. (This does not include harvesting and hauling which might add another $4–5/ton.)

Dehydration and separation costs were highest in the smallest model–$21.61 a ton. The largest model cost least–$11.31 a ton. Separation added between $1.70–$3.24 a ton over the standard model costs in the smallest group. In the largest volume group, separation increased the cost between $0.30 to $1.03 a ton.

Because dehydrated alfalfa is unstable under ordinary storage conditions, alfalfa dehydrators increasingly use inert gas storage to preserve product quality. Storage costs, including those for inert gas, for standard models ranged from $7.49 a ton in the smallest model to $5.40 in the largest. Models separating alfalfa had slightly higher costs for additional storage facilities and conveying equipment–$7.69 for the smallest and $5.69 for the largest.

Combining all costs allowed calculation of the total cost per ton. The most efficient separation model increased the total cost per ton between $1.81 to $0.47 over the standard model costs of $22.87 and $14.25. The highest cost separation model increased the per ton cost between $3.35 and $1.20.

Once having the production costs, the next question is the value of the products. During the course of the investigations, we prepared materials for several feeding trials with sheep and cattle. Hibbs *et al.* (1969) at Nebraska fed late first cutting and early third cutting unground stem-enriched alfalfa to growing cattle as a sole source of feed and to finishing lambs where the alfalfa product was blended 50:50 with ground shell corn. In both experiments, the stem products yielded performance results equal to or better than the chopped whole alfalfa hay. In the beef experiment the third cutting stem fraction, which was of relatively high quality (e.g., 15 percent protein), produced higher gains than all other rations including the all concentrate ration.

At Illinois, Tiwari and Garrigus (1971) tested alfalfa stem meal (about 17 percent protein) for growth performance at levels of 5 and 10 percent in a urea-containing high-concentrate diet comparing it to a standard type alfalfa meal (17 percent protein). The gains and conversions of lambs on the stem meal were numerically greater than those of lambs on the standard alfalfa, but the differences were not significant. Since the two products had the same protein content, one might expect the "stem meal" to be superior since it was prepared from alfalfa with a higher protein content with presumably greater digestibility coefficients for protein and fiber.

At Purdue (Beeson *et al.*, 1969), 15 percent protein alfalfa "stem" fraction was compared with 17 percent protein standard-grade dehydrated alfalfa in the Purdue 64 dry high urea supplement. Gains and feed conversions were essentially equal.

At Kansas State (Stiles *et al.*, 1968), a low quality stem meal (about 13 percent protein) was compared with other roughages in a hay–grain pellet-type dairy calf starter. In this experiment there was no significant difference between the dehydrated alfalfa control, alfalfa hay, the dehydrated stem meal fraction or corn cobs in the complete pelleted ration.

Additional studies have been carried out on composition and nutrient availability in the leaf fraction. The results are being utilized in an overall economic evaluation of dry leaf separation with the aid of parametric linear programming to determine value of the products for different uses in various market areas.

Separation of High- and Low-Protein Products by Dual Cutting Systems

Another system of separating protein-rich fractions from fibrous portions of alfalfa consists of cutting and collecting the top part of the plant, to segregate a protein-rich fraction, and harvesting the bottoms of the plants separately to obtain the low-protein, high-fiber fraction. This approach has been called dual cutting, double cutting, or "storeyed" cutting. The leaf fraction might then be dehydrated, while the bottom fraction might be sun-cured, ensiled, or possibly fed as green chop. In this country, considerable work has been done by Ogden and Kehr (1968) at the Nebraska station. Some earlier work had been done in southern Europe (Klimes and Verosta, 1962).

Several years ago, when we had our pilot plant in operation in Nebraska, we set up a cooperative experiment with Ogden *et al.* (1969) to determine the relative effectiveness of the dual-cutting approach as compared with dry separation and then to see whether a combination might have advantages over either by itself.

TABLE 8 Comparison of Two Harvesting and Fractionating Systems[a]

	Dry-Separation System			Dual-Cutting System	
	Whole	Fine	Coarse	Top Half	Bottom Half
Weight (%)	100.0	48.6	51.4	40.2	59.8
Protein (%)	20.7	25.3	16.3	25.7	17.0
Fiber (%)	27.5	20.6	34.6	18.9	31.5
Xanthophyll (mg/lb)	134.7	167.6	103.6	144.5	102.1

[a]Eight percent bloom alfalfa at feed rate of 1,925 lb/hr.

The data in Table 8 show that the dry separation procedure and dual cutting can yield comparable products, although a somewhat larger yield of fine fraction was obtained by dry separation. When the tops and bottoms obtained by double harvesting were dehydrated and then air separated, a substantial yield (about 36 percent) of a supergrade product was obtained that ran almost 30 percent protein and only 15 percent fiber (Table 9). The "stem fraction" from the bottom half of the plant ran about 15 percent protein and 35 percent fiber. The "overs" from the air separation of the bottom of the plants could be combined with the "unders" from the tops of the plants to produce a large yield of product containing about 22 percent protein and a little less than 24 percent fiber.

The practicality of the dual-cutting system has not been proven. It appears that the main problems revolve around handling the lower parts of the plant after leaf harvest. The farmer is interested in getting his crop out of the way and letting the next crop come on without wheeling heavy equipment over the field a second time. A possible solution would be to design a harvester that would cut the tops and the bottoms and segregate them to separate trailers. The leaf would

TABLE 9 Comparison of Tops and Bottoms Obtained by Dual-Cutting System[a]

	Top Half[b]			Bottom Half[b]		
	Whole	Fine	Coarse	Whole	Fine	Coarse
Weight (%)	100.0	36.1	63.9	100.0	34.3	65.7
Protein (%)	25.7	29.7	23.4	17.0	21.0	14.9
Fiber (%)	18.9	15.1	21.1	31.5	24.2	35.4
Xanthophyll (mg/lb)	144.5	147.7	142.8	102.1	124.6	90.4

[a]Eight percent bloom alfalfa.
[b]Feed rate for top half = 2,450 lb/hr; for bottom half = 3,100 lb/hr.

then go to the dehydrator, while the bottoms would go into silos or directly to the feedlots. Dual cutting has not yet been used commercially, and it will be interesting to see if and how it will fit into the evolving processing industry.

Fractionation of Alfalfa by Separation of Cell Contents from Cell Walls

It has long been known that the contents of the cells of either leaves or stems are essentially fiber free, while the cell walls are fibrous. The most obvious means of separating the cell contents from the fiber is by pressing out the juice. The juice could either be dried per se to yield a 35 percent protein product or heat-coagulated to yield a 45–50 percent protein concentrate. This is an old idea. In the 1930's and 1940's, work on it was going on at the Western Regional Laboratory, Nebraska, at several locations in England, in Hungary, and in many other places. At Rothamsted in England, N. W. Pirie has worked in the field continuously for over 30 years (Pirie, 1957, 1966, 1969). Several U.S. and English companies went as far as pilot plant scale and supplied test products to many universities and experiment stations. What happened to these efforts?

Even though there were still some unsolved technological problems in the 1940's and 1950's, we could do a reasonably good job of producing high-protein products (Bickoff *et al.*, 1947; Crook, 1946; Kohler and Graham, 1951). However, no one was able to put the economic and marketing factors together in a way that would justify the expenditures necessary to build a plant and develop markets for the products in the feed industry. Conversion of the juice or protein isolate from it to palatable foods was not solved, even from a technological standpoint.

What has happened during the past several decades to alter the probability of success of a juicing operation? First, computer studies show that for poultry rations, using a low-energy feedstuff in a ration is expensive. If dehydrated alfalfa is used to supply a specific factor such as xanthophyll, the energy deficit incurred must be made up by adding costly concentrated energy sources. The higher the energy of the formulated ration, the greater the penalty the alfalfa meal takes for its energy deficit. Hence, there has been a scramble to find other xanthophyll sources, and alfalfa products low in fiber are needed to meet these market demands.

Second, there is a growing need for long-fiber roughage in dense form for cattle, both dairy and beef. Protein content is becoming of less importance in cattle feed ingredients, since increasing use is being made of low-cost urea and other NPN products in place of protein.

Third, insertion of a juicing operation ahead of the dehydrator results in a 30–50 percent increase in dehydrator throughput. This results from reduced moisture content and crushing of the difficult-to-dry stems. The bulk of the water removed by pressing can then be evaporated in a multiple-effect evaporator that requires much less energy per pound of water evaporated than a conventional dehydrator. In this connection, it is important to note that removal of up to 40 percent of the water of alfalfa by pressing has only a slight effect on the quality of the dehydrated alfalfa produced. Removal of nutrients in the juice is compensated in part by lesser losses in the dehydrator due to lower retention time.

Finally, by parametric linear programming, the potential value of a new product can be fairly accurately determined, thus lessening the risk involved in its commercial development.

About 5 years ago, we re-entered this field of research in collaboration with a company, Batley-Janss Enterprises (BJE), Brawley, California, which is in the large-scale production of dehydrated alfalfa (Harley, 1970). In accord with the considerations above, we evolved a program, the immediate (Phase I) objectives of which were to produce superior animal feeds (Kohler *et al.*, 1968). We felt the success of this Phase I effort would serve as an economically sound base for a Phase II development of low-cost leaf protein concentrate (LPC) for human food.

The Phase I specific research objectives were (a) reduction of costs of dehydrated alfalfa by mechanical dewatering, (b) recovery of a high-protein–high-xanthophyll product (PRO-XAN) designed primarily as a pigmentation supplement for poultry, and (c) recovery of the solubles fraction from the coagulation step as a molasses or unidentified growth factor concentrate (Spencer *et al.*, 1970, 1971; Lazer *et al.*, 1971). Once a high-value market was established for PRO-XAN, it was anticipated that more vigorous pressing might be used to increase yields. Then, of course, a lower-value dehydrated product would result, competitive with alfalfa hay.

Figure 2 is a schematic diagram showing the relation of conventional dehydration (upper part of figure) to a process including dry leaf–stem separation (center of figure) and to the Phase I wet separation process (lower line of figure). The latter process has been dubbed the PRO-XAN process. The name PRO-XAN is also used to designate

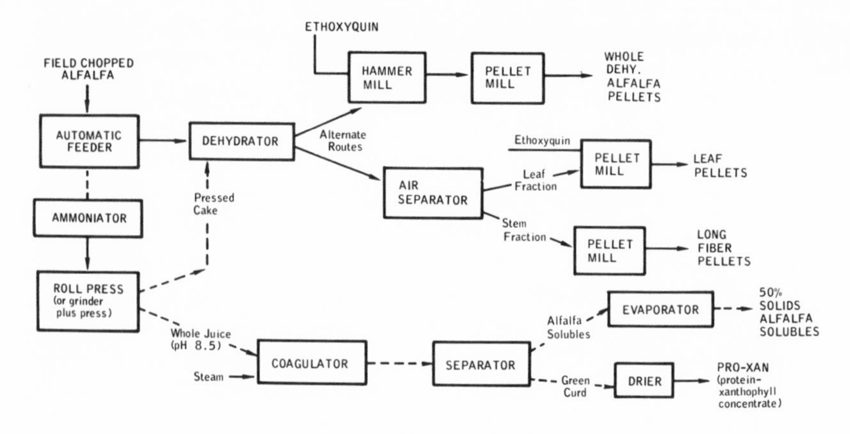

FIGURE 2 The PRO-XAN process.

the protein–xanthophyll concentrate derived from the process as a by-product of the dehydrated alfalfa main product. The juicing and coagulation steps are derived from the earlier work mentioned, but the unit process variables had to be studied to achieve the objectives of obtaining a high-quality dehydrated product and minimum xanthophyll and protein losses.

Since we were not initially interested in maximum yields of protein in the juice fraction but rather in large throughput and maintaining quality of the pressed cake to produce a high-grade alfalfa meal, we chose a sugar cane roll press to replace both grinding and pressing operations recommended by some investigators with other objectives in mind. An important consideration was that sugar processing equipment is available for very large-scale operations and would require little or no machine research and development. In our pilot plant rolls we can obtain 35–50 percent of the weight of the fresh young alfalfa as press juice at a feed rate of about 1,000 lb/hr. Figure 3 shows a balance sheet on solids, protein, and water in a 35 percent juice removal operation (de Fremery, 1971). Alfalfa that normally would give a 20 percent protein dehydrated product yielded a dehydrated press residue containing 19.6 percent protein. You will note that only 4½ percent of this raw material is recovered as PRO-XAN and 87 percent as high-grade alfalfa meal. A small shift in price of the dehydrated would obviously overwhelm our profit or loss from the PRO-XAN product. Also, note that there is 3½ times as much brown juice solids removed as PRO-XAN. The brown juice solids are low in crude protein derived from alfalfa and high in ash and carbohydrates. This accounts in part for the relatively high protein in pressed dehydrated alfalfa. Some water-soluble vitamins are removed, but these are so cheap that practically no computer programs include them except as synthetic additives. More importantly, water-soluble saponins are removed in the process, a positive effect. Table 10 shows that dehydrator throughput was increased by about 40 percent in a typical experiment, thus greatly reducing the cost of dehydrated alfalfa. The pressed dehydrated alfalfa could still be air-classified to produce a 25 percent protein meal if desired for use in rations of monogastric animals.

Initially, we were dismayed to find that large amounts of xanthophyll were lost during and immediately following rolling. The importance of xanthophyll losses is illustrated by the fact that computer evaluations of projected products show that the xanthophyll in a low-fiber product is worth 7 to 15¢/g, depending on types of poultry rations being considered and prices of competitive xanthophyll-containing feeds including alfalfa meal. Thus, a PRO-XAN with 40 percent protein and 700 mg/lb xanthophyll would have been

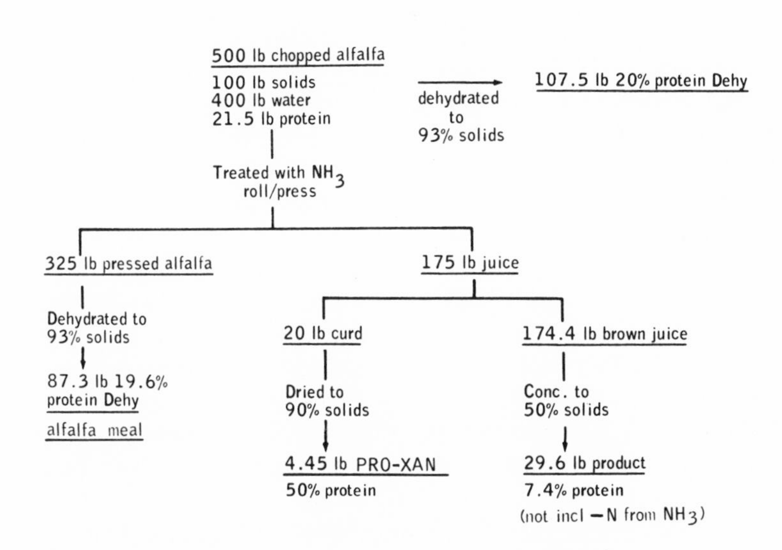

FIGURE 3 Material balance through PRO-XAN process at 35 percent juice removal level.

TABLE 10 Comparison of Dehydration Rate and Separability of Whole Alfalfa and Dewatered Alfalfa

	Dehydration Rate (lb/)	Leaf Separated (%)	Protein	Xanthophyll
Whole dehydrated alfalfa	335	–	22.8	202
Leaf fraction	–	42.3	25.1	214
Dewatered, dehydrated alfalfa	446 (+39%)	–	21.3	183
Leaf fraction	–	42.0	24.5	204

worth over $300/ton in a layer ration based on the Kansas City, December 1966 market.

We found that by adjusting the pH of the alfalfa to 8 or above, the losses of xanthophyll were reduced presumably due to the low activity of lipoxidase at more alkaline pH's. It causes the curd formed on steam injection to be harder and more readily handled. It keeps the product green by preventing conversion of the bright green chlorophyll to olive brown pheophytin. This is desirable in feed products. Although it has not been clearly demonstrated, we feel that the yield of protein should be increased by ammonia addition, since other bases have been shown to have such an effect. Finally, the addition of ammonia would be expected to reduce protein losses through autolysis during handling and extraction. This is suggested by the work of Singh (1962), who found that protease enzyme system of wheat leaves shows maximal activity at about pH 5.5 and is greatly reduced at pH 8 and over.

As our laboratory and pilot plant work progressed, our industry collaborator moved ahead installing full-scale equipment approximately 50 times as large as ours.

The Imperial Valley plant is presently operating on a 24-hr basis and all of the new product (X-PRO® brand of PRO-XAN) is being sold, some domestically and some abroad. The brown juice, which in our pilot plant is being recovered as a 50 percent solids alfalfa solubles concentrate, is not yet being so recovered in the BJE plant. Instead, it is being returned to the alfalfa prior to dehydration and thus is recovered in the dehydrated alfalfa product. Consequently, the full potential of the process is not being utilized commercially at present. Our current research is directed to (a) establishing uses for the alfalfa solubles, (b) improving the unit processes of the Phase I PRO-XAN process including work on xanthophyll stability, pressing efficiency, double rolling, other grinders and rollers, curd dewatering, product uses, (c) studying the economics of the Phase I process in collaboration with the Economics Research Service to estimate costs of production at various locations and with varying levels of extraction, and (d) Phase II developing of a separation process to produce edible-grade protein that is free of color and palatable from the protein mixture now recovered as PRO-XAN.

REFERENCES

Ayres, G. E. 1968. Field harvesting alfalfa leaves, p. 97–103. ARS-74-46, Western Regional Research Laboratory, U.S. Department of Agriculture, Berkeley, Calif.

Beeson, M. W., T. W. Perry, C. F. Hatch, and M. T. Mohler. 1969. Nutritional factors affecting the utilization of dry and liquid high-urea supplements, p. 35–39. *In* Annual Indiana cattle feeders day report. Purdue University, Lafayette, Ind.

Bickoff, E. M., A. Vevenue, and K. T. Williams. 1947. Alfalfa has a promising chemurgic future. Chem. Dig. 6:213, 216–218.

Chrisman, J., G. O. Kohler, A. C. Mottola, and J. W. Nelson. 1971. High and low protein fractions by separation milling of alfalfa. ARS-74-57, Western Regional Research Laboratory, U.S. Department of Agriculture, Berkeley, Calif.

Clarke, M. F. 1938. The nitrogen distribution in alfalfa hay cut at different stages of growth. Can. J. Res. Sec. C 16:339–346.

Crook, E. M. 1946. The extraction of nitrogenous materials from green leaves. Biochem. J. 40:197–209.

de Fremery, D. 1971. Yield and composition of fractions, p. 41–50. ARS-74-60. Western Regional Research Laboratory, U.S. Department of Agriculture, Berkeley, Calif.

Harley, R. 1970. Is alfalfa our next high-protein source? Farm Q. July–Aug: 40–41, 70–71.

Hibbs, J. R., T. J. Klopfenstein, and T. H. Doane. 1969. Quality of alfalfa stems. J. Anim. Sci. 27:1768.

Klimes, I., and V. Bohuslar. 1962. Neve Mlethode der Erzevgung von Eiweiss-Vitamin Konzentrat aus Futterpflanzen. Arch. Tierenaehr. 11:393.

Kohler, G. O., and W. R. Graham, Jr. 1951. A chick growth factor found in leafy green vegetation. Poult. Sci. 30:484–491.

Kohler, G. O., and J. Chrisman. 1966. Separation milling of alfalfa, p. 46–58, CR-58-66, Agricultural Research Service, U.S. Department of Agriculture, Beltsville, Md.

Kohler, G. O., E. M. Bickoff, R. R. Spencer, S. C. Witt, and B. E. Knuckles. 1968. Wet processing of alfalfa for animal feed products, p. 71–79. ARS-74-46, Western Regional Research Laboratory, U.S. Department of Agriculture, Berkeley, Calif.

Lazar, M. E., R. R. Spencer, B. E. Knuckles, and E. M. Bickoff. 1971. PRO-XAN process: Pilot plant for separation of heat-precipitated leaf protein from residual alfalfa juice. Agric. Food Chem. 19(5):944–946.

Nageli, C. 1899. Cited in A. Guilliermond, 1941. The cytoplasm of the plant cell. Chronica Botanica, Waltham, Mass.

Ogden, R. L., and W. R. Kehr. 1968. Field management for dehydration and hay production, p. 23–37. ARS-74-46, Western Regional Research Laboratory, U.S. Department of Agriculture, Berkeley, Calif.

Ogden, R. W., W. R. Kehr, J. Chrisman, and G. O. Kohler. 1969. 1968 cooperative field-management air-separation alfalfa project at Lexington, Nebraska. Reported at the Nebraska Alfalfa Dehydrators Association Meeting, Nebraska Center, Lincoln, Jan. 1969.

Pirie, N. W. 1957. Leaf protein as human food. Food Manuf. 32:416–419.

Pirie, N. W. 1966. Leaf protein as human food. Science 152:1701–1705.

Pirie, N. W. 1969. The production and use of leaf protein. Proc. Nutr. Soc. 28:85–91.

Shepherd, J. B., H. G. Wiseman, R. E. Ely, C. G. Klein, W. J. Sweetman, C. H. Gordon, L. G. Schoenleber, R. E. Wagner, L. E. Campbell, G. D. Roane, and W. H. Hostermana. 1954. Experiments in harvesting and preserving alfalfa for dairy cattle feed. U.S. Department of Agriculture, Agricultural Research Service, and Agric. Mark. Serv. Tech. Bull. 1079.

Singh, N. 1962. Proteolytic activity of leaf extracts. J. Sci. Agric. 13:325–332.

Spencer, R. R., E. M. Bickoff, G. O. Kohler, S. C. Witt, B. E. Knuckles, and A. C. Mottola. 1970. Alfalfa products by wet fractionation. Trans. Am. Soc. Agric. Eng. 13(2):198–200.

Spencer, R. R., A. C. Mottola, E. M. Bickoff, J. P. Clark, and G. O. Kohler. 1971. The PRO–XAN process: The design and evaluation of a pilot plant system for the coagulation and separation of the leaf protein from alfalfa juice. Agric. Food Chem. 19(3):504–507.

Stiles, D. A., E. E. Bartley, A. D. Dayton, and C. W. Deyoe. 1968. Nutritive value of various roughages used in dairy calf starters as measured by calf growth. Report of Progress 141, Animal Science Progress Report for 1968. S. E. Kansas Branch Sta. Kansas State University, Manhattan, Kan.

Tiwari, A. D., and U. S. Garrigus. 1971. Utilization of air-separated stem fraction of dehydrated alfalfa meal of lambs. J. Anim. Sci. 33(4):903–905.

Vosloh, C. J. Jr. 1970. Alfalfa dehydration, separation and storage: Costs and capital requirements. Marketing Res. Rep. No. 881. U.S. Department of Agriculture, Economic Research Service, Washington, D.C.

Whitney, L. F., and C. W. Hall. 1963. Harvesting and drying of alfalfa leaves. Proc. Winter Meeting, American Society of Agricultural Engineers, St. Joseph, Mich.

D. M. Doty

DEVELOPMENTS IN PROCESSING MEAT AND BLOOD BY-PRODUCTS

INTRODUCTION

It is essential that animal protein by-products be more effectively utilized than has been the case in the past. This is necessary because of the contribution that animal by-products makes to total pollution and because the current world protein shortage could be alleviated in part by better processing and improved utilization of animal by-product proteins. For best utilization of these products, we need additional information on the composition and nutritive quality of these materials as they relate to different methods of processing.

BLOOD PROTEIN

Dried blood has been used for animal feeds and adhesives for many years. The adhesive market has been replaced by materials that are more water repellent; this has made larger amounts of dried blood available for feed. Unfortunately, the conventional methods used for drying blood give a product in which some of the amino acids, particularly lysine, are not readily available to the animal.

TABLE 1 Amino Acids (%) in Blood Dried by Different Processes[a]

Amino Acid	Conventional	Spray Dried	Ring Dried
Isoleucine	0.9	0.9	1.0
Leucine	11.9	11.7	12.7
Lysine	8.0	8.5	9.1
Methionine	1.3	1.4	1.4
Phenylalanine	6.8	6.6	7.2
Threonine	4.4	4.1	4.6
Valine	8.3	8.2	9.0

[a]From P. E. Waibel and R. J. Meade (personal communication).

Recent studies at the University of Minnesota (P. E. Waibel and R. J. Meade, personal communication) have shown that the overall amino acid composition of blood dried by different methods is very similar (Table 1). However, the available lysine as determined by bioassay differs greatly in blood that has been spray dried or ring dried as compared with blood dried by conventional means (Table 2). It is likely also that the nutritional availability of other essential amino acids will be adversely affected by conventional methods of drying.

Even if most of the animal blood available were spray dried or ring dried, as these processes are now used, there would still be a sizable loss of blood protein because about half of the serum protein is removed from the coagulated blood before drying. Ideally, the serum protein, as well as the red cell protein, should be recovered, preferably without the presence of the heme pigment.

Studies that have accomplished this have been performed by researchers at Texas A&M University (W. A. Landmann and C. W. Dill, personal communication). It has been possible to isolate both the serum protein fraction and a red cell protein fraction from whole blood as shown in Figure 1. The two protein fractions can be

TABLE 2 Biological Lysine Availability (%) of Dried Blood (Turkeys)[a]

Conventional	Spray Dried	Ring Dried
42 ± 9.8	44 ± 11.0	86 ± 14.0
39 ± 10.6	81 ± 14.3	87 ± 6.0
28 ± 13.9	83 ± 19.0	82 ± 10.2

[a]From P. E. Waibel and R. J. Meade (personal communication).

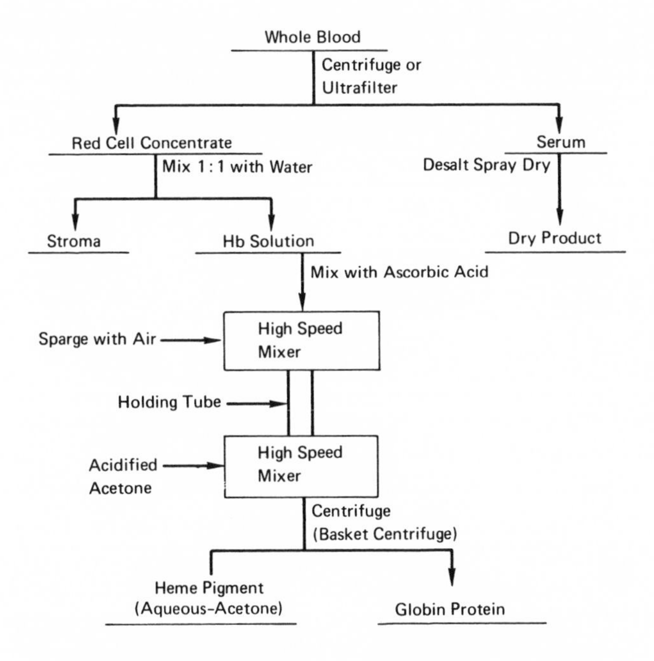

FIGURE 1 Scheme for separation of protein fractions from animal blood. From W. A. Landmann and C. E. Dill (personal communication).

recovered as dry, free-flowing, essentially colorless powders that should find use in human food as well as livestock feed. The essential amino acid composition of the two protein fractions show that these protein powders are excellent sources of lysine and tryptophan (Table 3). Also these two protein fractions have excellent

TABLE 3 Essential Amino Acids (% of Protein) in Bovine Blood[a]

	Whole Blood	Plasma	Hemoglobin (Decolorized)
Isoleucine	1.1	3.2	–
Leucine	12.5	9.0	14.2
Lysine	11.4	9.5	10.9
Methionine	1.5	1.2	1.8
Phenylalanine	6.6	5.0	8.1
Threonine	4.9	6.2	4.4
Tryptophan	–	2.0	0.8
Valine	8.3	6.6	9.6

[a]From W. A. Landmann and C. W. Dill (personal communication).

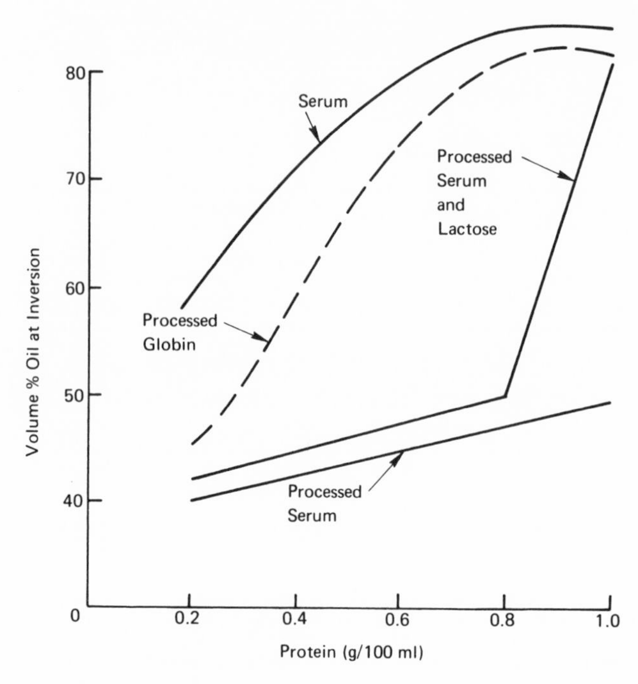

FIGURE 2 Emulsifying capacity of blood protein fractions. From Tybor *et al.* (in press).

functional characteristics (Tybor *et al.*, in press), especially the ability to emulsify large quantities of fat (Figure 2). This would be particularly important for food use and might also be of value if these materials were used in milk replacers for young animals.

The samples that have been produced in a laboratory pilot unit have excellent microbiological characteristics. Total counts have generally been less than 2,000 microorganisms/g. No *Salmonella, Clostridium,* or *E. coli* have been detected in any of the samples. Nutritional evaluation of the two protein products is in progress.

MEAT AND BONE MEAL

Meat and bone meal, as it is now produced, is somewhat variable in composition and in nutritive value because of variations in the raw materials. Despite this, however, the composition (Doty, 1969) at a given protein level is fairly uniform (Table 4). The amino acid com-

TABLE 4 Proximate Analysis of Animal By-Product Meals[a]

	45% Protein	50% Protein	53–55% Protein	60% Tankage
Protein	46.5 ± 2.8	51.6 ± 2.3	54.4 ± 2.8	61.7 ± 2.6
Moisture	5.3 ± 1.6	5.2 ± 1.2	5.4 ± 2.0	6.9 ± 3.2
Fat	10.3 ± 3.4	10.1 ± 3.1	8.8 ± 3.5	7.5 ± 2.3
Ash	35.5 ± 4.1	28.7 ± 3.6	27.5 ± 4.2	19.2 ± 4.8

[a]From Doty (1969).

position, of course, varies with protein level because the lower protein materials are derived largely from collagens. Material containing higher amounts of protein is that from raw material with a desirable amino acid balance (Table 5). Meade (1969) found that one third to one half of the supplementary protein in corn–soybean meal rations for growing swine could be furnished by meat and bone meal. Runnels (1968) reported that at least 10 percent meat and bone meal could be used in least-cost formulated broiler rations that gave maximum growth and feed efficiency. Burgos *et al.* (1972) found that amino acids in meat and bone meal were more than 95 percent available to broiler chicks.

More effective utilization of animal by-product meal could be achieved, however, if the high-quality protein could be separated from collagen-type proteins, since animal by-product meals, *in toto*, include about 50 percent of the latter. Consequently, improvements in nutritive value might be accomplished by (1) segregation of raw material, (2) separation of different types of protein during process-

TABLE 5 Amino Acids (%) in Animal By-Product Meals[a]

(%)	45% Protein	50% Protein	60% Tankage
Hydroxyproline	3.8	3.2	1.7
Isoleucine	1.4	1.4	1.1
Leucine	2.4	2.9	5.9
Lysine	2.3	2.4	4.4
Methionine	0.6	0.7	0.7
Phenylalanine	1.4	1.6	3.2
Threonine	1.3	1.6	2.3
Tryptophan	0.2	0.3	0.7
Valine	1.7	2.0	1.1

[a]From Doty (1969).

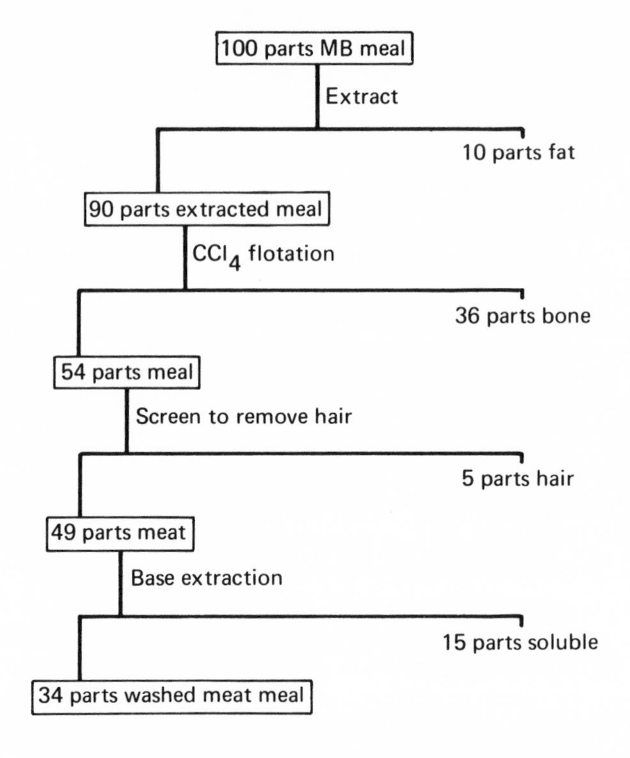

FIGURE 3 Scheme for isolation of high quality protein from meat and bone meal. From Nash and Mathews (1971).

ing, or (3) conversion of collagen to a protein of higher nutritive quality.

Olson (1971) and Levin (1971) reported that high-quality protein powders could be prepared from poultry offal by azeotropic extraction. Nash and Mathews (1971) were able to isolate a high-quality protein meal from meat and bone meal by a series of extractions and mechnical separations (Figure 3). The final product was bland and nearly colorless but only slightly soluble in water. It had a good amino acid profile (Table 6) and exhibited high nutritive quality in rat-feeding tests (Figure 4).

However, the yield of the final product was only about 34 percent of the starting meat and bone meal, and the estimated net product cost would be about 18¢/lb (90 percent protein).

Several years ago the Fats and Proteins Research Foundation (FPRF) sponsored research on the development of an enzymatic-

TABLE 6 Essential Amino Acids in High-Protein Meat and Bone Meal Fraction[a] (Grams per 16 Grams N)

Isoleucine	3.3	4.2
Leucine	7.0	4.8
Lysine	5.5	4.2
Methionine	1.6	2.2
Phenylalanine	3.6	2.8
Threonine	3.4	2.8
Tryptophan	0.9	1.4
Valine	5.0	4.2

[a]From Nash and Mathews (1971).

rendering system that would yield a soluble protein powder of high nutritive quality, a bone fraction, and a collagen fraction. The process was developed (Hayhow and Flinn, 1967) and the amino acid profile of the soluble protein powder was superior to that of meat and bone meal (Table 7).

FERMENTATION OF COLLAGEN

The fractionation process would not be economically feasible unless the collagen fraction can be supplemented or upgraded to a protein

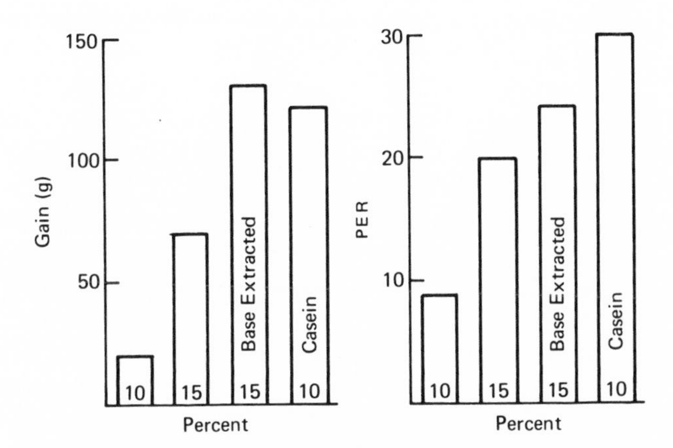

FIGURE 4 Gain and protein efficiency ratios of rats fed protein fractions from meat and bone meal. From Nash and Mathews (1971).

TABLE 7 Amino Acids (%) in Conventional and Enzyme-Hydrolyzed Meat Meal[a]

(%)	Conventional	Enzyme Hydrolyzed
Protein	51.6	78.0
Hydroxyproline	3.2	2.7
Isoleucine	1.4	2.8
Leucine	2.9	5.3
Lysine	2.4	3.4
Methionine	0.7	1.2
Phenylalanine	1.6	2.3
Threonine	1.6	2.3
Tryptophan	0.3	0.4
Valine	2.0	3.8

[a]From Hayhow and Flinn (1967).

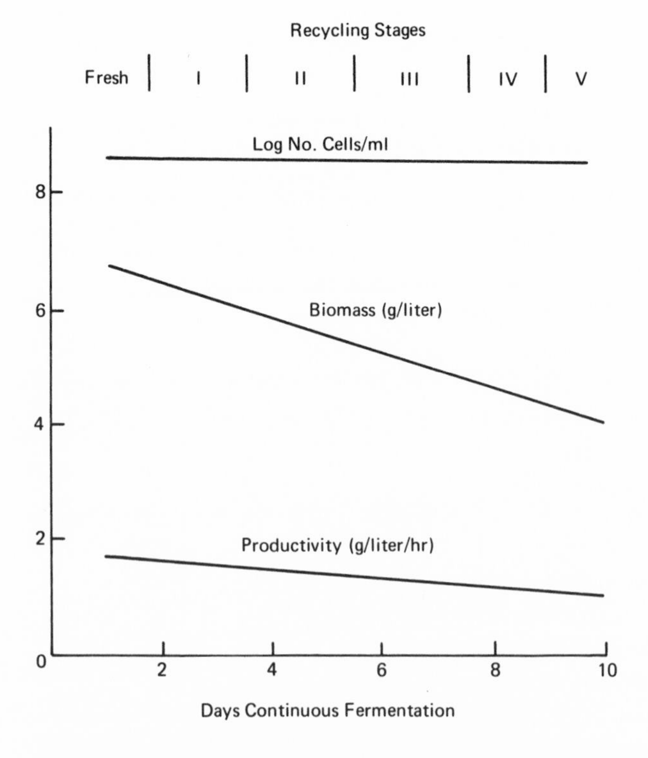

FIGURE 5 Biomass and productivity of *B. megaterium* grown on solubilized collagen through five cycles in a continuous fermentation procedure. From Bough *et al.* (1972).

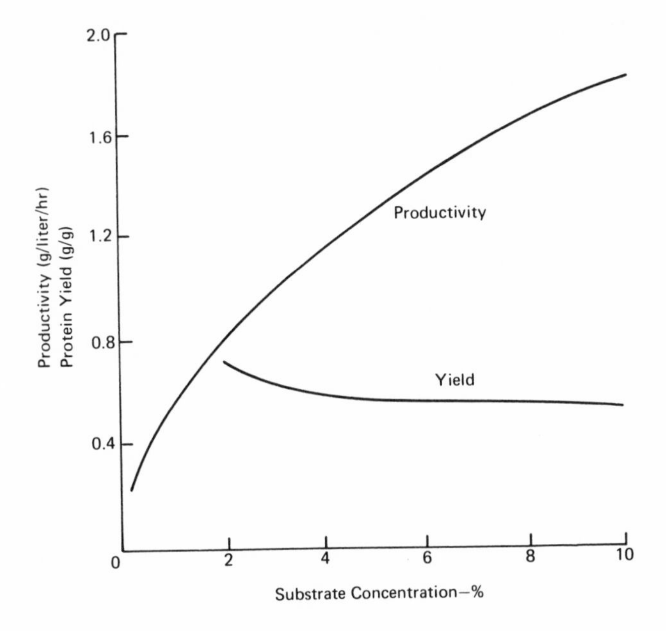

FIGURE 6 Influence of substrate concentration on productivity and yield of *B. megaterium* grown on solubilized collagen. From Bough *et al.* (1972).

of higher nutritive quality. This consideration led to extensive investigation of the microbial conversion of collagen to single-cell protein (Bough *et al.*, 1972).

In these studies it was found that *Bacillus megaterium* grows profusely on media containing solubilized collagen as the sole source of nitrogen and energy. More complete conversion of collagen protein to microbial protein (yield) can be obtained if glucose or other suitable carbon source is used as a supplementary energy source and if the pH is maintained near the neutral point by the resulting formation of lactic acid. With continuous aerobic fermentation at a dilution rate of 0.25/hr and glucose addition to control pH, the cell population is stable at approximately 10^8 cells/ml (Figure 5). Productivity increases with substrate composition to a level of approximately 1.8 g/liter/hr at a substrate concentration of 10 percent (Figure 6). However, protein yield drops unless a minimal supplement is included in the fermentation medium. The bacterial protein has an excellent amino acid profile (Table 8) and has a PER value about 78 percent that of casein.

TABLE 8 Essential Amino Acids in Substrate and Bacterial Protein[a] (Grams per 16 Grams N)

	SP-100	*B. megaterium*
Isoleucine	2.1	4.4
Leucine	3.6	11.0
Lysine	4.3	6.3
Methionine	1.3	3.1
Phenylalanine	0.9	5.4
Threonine	2.4	4.8
Tryptophan	0.2	–
Valine	3.1	6.5

[a]From Bough *et al.* (1972).

At current price levels and protein demand in the United States, it would probably not be economically feasible to convert collagen to single-cell protein. However, the information available indicates that this conversion would be just as economical as is the conversion of petroleum fractions to single-cell protein.

HAIR AND FEATHERS

The industrial use of hair and feathers has gradually decreased, and the only sizable potential market for these keratin proteins is as ingredients in poultry feeds. Hydrolyzed feather meal has been used in poultry rations for many years, but its value has been questioned by some nutritionists because of its relatively low and variable digestibility as determined by the *in vitro* pepsin digestibility method. However, recent studies by Professor E. L. Stephenson, University of Arkansas, have shown that the amino acids in hydrolyzed feather meal are more than 95 percent utilized by the chick. Professor Stanley Balloun, Iowa State University, found that the different processing conditions used commercially had little or no influence on the amino acid availability in broiler rations.

Other investigators have reported that the amino acids in properly hydrolyzed hog hair are completely available to the chick. Studies performed for FPRF by Harris Research Laboratories (now a part of Gillette Research Institute) showed that the amino acids in hair or feathers could be made completely available to the chick by appropriate chemical treatment to split the –S–S– bonds in the keratin molecule. Thus the basic information is available to allow feed manufacturers to incorporate properly processed hair and feather meal into poultry rations. Available hair and feathers could be recovered and utilized in this way if appropriate modifications were made in the various state regulations governing the sale of feeds and feed ingredients.

SUMMARY

Much of the animal by-product protein material that is now incompletely recovered or poorly utilized could be processed and used in human food or animal feeds. Proteins from animal blood can be produced in highly purified form. These blood protein preparations have excellent nutritional and physical characteristics. With appropriate processing, animal offal proteins and a high-protein fraction from meat and bone meal of high nutritive quality can be produced.

Collagenous proteins, which contain limited amounts of the essential amino acids, can be converted to a single-cell protein of high nutritive quality by fermentation.

Keratin proteins (hair and feathers) can be processed to make their constituent amino acids almost completely available to poultry. These proteins can be recovered and utilized entirely as ingredients in poultry rations.

REFERENCES

Bough, W. A., W. L. Brown, J. D. Porsche, and D. M. Doty. 1972. Utilization of collagenous by-products from the meat packing industry: Production of single-cell protein by continuous cultivation of *Bacillus megaterium*. Appl. Microbiol. 24:226–235.

Burgos, A., J. I. Floyd, and E. L. Stephenson. 1972. Amino acid content and availability of different meat and bone meal samples in the broiler chick. Feedstuffs 44(16):15.

Doty, D. M. 1969. Nutritional constituents in animal and poultry by-product meals. Feedstuffs 41(11):24.

Hayhow, W. R., and J. E. Flinn. 1967. Development of the selective enzymatic rendering process. Report from Battelle Memorial Institute to Fats and Proteins Research Foundation, Dec. 1967. 23 p. (unpublished)

Levin, E. 1971. Conversion of meat by-products into edible protein concentrates, p. 29–38. *In* Proceedings of the Meat Industry Research Conference, Mar. 26–27, 1970. American Meat Institute Foundation.

Meade, R. J. 1969. Animal by-products as sources of amino acids for growing swine. Feedstuffs 41(11):43.

Nash, H. E., and R. J. Mathews. 1971. Food protein from meat and bone meal. J. Food Sci. 36:930–935.

Olson, F. C. 1971. Nutritional aspects of offal proteins, p. 23–28. *In* Proceedings of the Meat Industry Research Conference, Mar. 26–27, 1970. American Meat Institute Foundation.

Runnels, T. D. 1968. Meat and bone meal as an ingredient in broiler diets. Feedstuffs 40(42):27.

Tybor, P. T., C. W. Dill, and W. A. Landmann. The effect of decolorization and spray drying on the emulsification capacity of blood protein concentrates. J. Food Sci. (In press)

K. J. Smith

ADVANCES IN OILSEED PROTEIN UTILIZATION

Oilseed protein constitutes the largest source of supplemental protein in livestock rations. Hence, any discussion of increasing existing protein supplies must include oilseed proteins. This paper will discuss oilseed protein utilization, recent technological advances, and future utilization possibilities. In so doing, it will be apparent that oilseed protein supplies may be increased by two methods: increasing production of existing oilseed protein sources and changing utilization patterns so as to encourage the distribution of protein supplies to areas of greatest need.

Table 1 presents estimates of the quantity of high-protein feed supplements available for livestock production. These data are expressed on a 44 percent protein basis for uniformity. The data indicate that oilseed protein sources constitute three fourths of the presently available protein supplements. Animal proteins (tankage-meat meal, fish meal, and milk products) and grain proteins (brewers' and distillers' dried grains, gluten feed, and meal) comprise the balance of the 20 million tons of high-protein feed ingredients available for use in livestock rations.

TABLE 1 Quantity (in thousands of tons) of High-Protein Feed Available for Feeding[a,b]

	Oilseed Meal	Animal Protein	Grain Protein	Total
1969	15,310	3,444	1,321	20,075
1970[c]	15,227	3,533	1,319	20,079
1971[c]	15,000	3,640	1,360	20,000

[a] From Economic Research Service (1972).
[b] In 44 percent soybean meal equivalents.
[c] Preliminary estimates.

SOYBEAN MEAL

Table 2 provides data on oilseed meal production and export, showing that soybean meal makes up approximately 88 percent of the total oilseed meal produced in the United States. These data also indicate that approximately 25 percent of the soybean meal produced is exported. Cottonseed meal is the second largest oilseed protein source, followed by lesser amounts of linseed, peanut, and copra. The relative importance of soybean meal to the livestock producer is readily apparent from these data. Changes in soybean production and export will have a direct influence on livestock production.

There are several factors accounting for soybean meal domination of the supplemental protein sources. The demand for supplemental protein has given incentive to the production of soybeans as a cash crop. The existing need for supplemental protein simply creates a demand for soybeans and soybean meal that encourages greater

TABLE 2 Oilseed Meal Supply and Export (in thousands of tons)—United States[a,b]

	Domestic Use	Export	Total Production
Soybean	13,524	4,297	17,821
Cottonseed	1,743	23	1,766
Linseed	220	97	317
Peanut	148	–	148
Copra	91	–	91

[a] From Economic Research Service (1972).
[b] Mean estimate for 1969–1970.

production and domination. Then, too, the amino acid composition of soybean meal complements existing grain sources in rations for nonruminants. Grains will not meet the critical amino acid needs of the nonruminants, but simple rations may be formulated containing grain and soybean meal that will meet the animal needs and support economic performance.

The lack of plant toxins in properly prepared soybean meal has allowed for unlimited usage in livestock rations. Chemical compounds detrimental to animal performance are known to occur in oilseeds; however, present soybean processing knowledge essentially eliminates concern for these naturally occurring plant constituents.

If alternative sources of protein could be used to satisfy the ruminant protein needs, then greater amounts of soy protein would be available for nonruminants. Factors that increase production of soybeans have a direct influence on the availability of protein for livestock production. Increased soybean meal production could be utilized to increase supplies available for all markets or for selective end uses.

It is instructive to look back at changes in the soybean acreage harvested and to note how this increased production has satisfied protein needs (Figure 1). The Corn Belt States (Iowa, Illinois, Indiana, Ohio, and Missouri) increased their acreage from 13.8 to 21.9 million during the period 1960–1970. The greatest relative increase in acreage, from 3.5 to 8.3 million acres during the 10-year period, was in the Delta States (Arkansas, Missouri, and Louisiana). The total acreage in the United States increased from 23.7 to 42.5 million acres.

Figure 2 presents soybean production and consumption data from 1964 through 1970. The increased acreage planted to soybeans has been responsible for greater soybean production, since soybean yields per acre have not increased significantly during the past decade. There has been much talk about potential soybean yield breakthroughs; however, agronomic improvements to date and grower acceptance of soybean varieties that are genetically higher in protein have not resulted in improved protein yields. The point is that if these genetic breakthroughs become realities, then the role of soybeans may become even more dominant in supplying the world a protein source. Increased yields of soy protein per acre could do much to satisfy the world's need for a relatively cheap protein source.

One way of enlarging available protein supplies is by continuing to increase the acreage planted to soybeans. To do so involves gov-

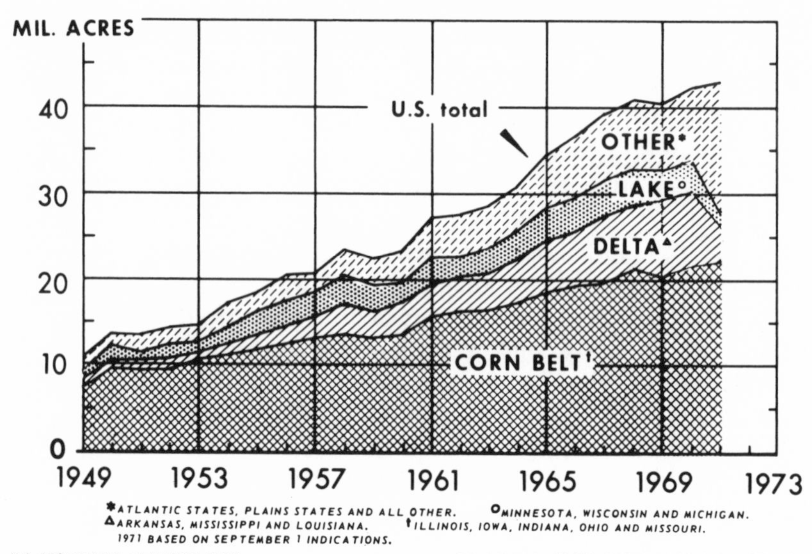

FIGURE 1 Soybean acreage harvested.

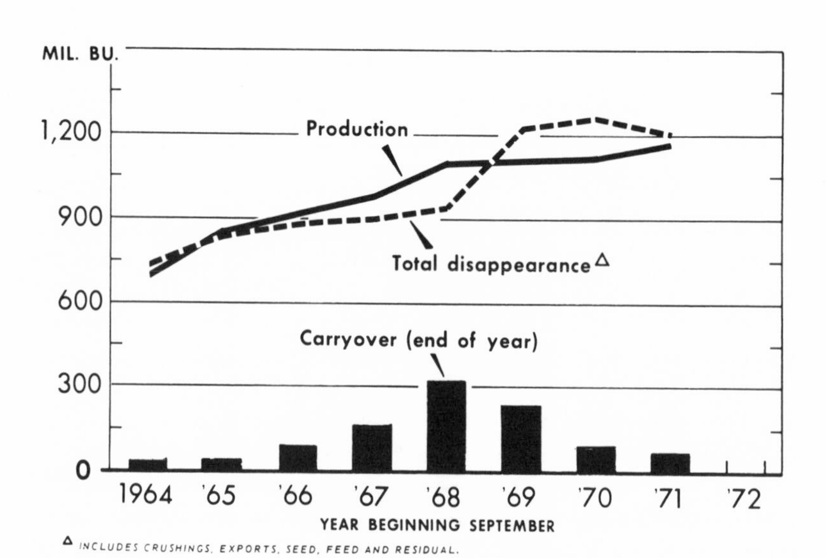

FIGURE 2 Soybean output and carry-over.

ernment programs, alternative crops, management decisions, etc. Favorable economic returns to the grower could increase domestic protein supplies sginficantly if needed.

Another source of soybeans for domestic use could be obtained by changes in soybean exports. It is recognized that the area of world trade is complex; however, the United States is exporting oilseed protein. If the domestic demand became sufficient, withholding oilseeds from the export market could augment the total protein available to the United States.

COTTONSEED MEAL

Unlike soybean meal, which depends on supply alone, cottonseed meal is affected both by supply and utilization decisions. Cottonseed is a by-product; the amount available depends upon cotton production. Therefore, factors affecting cotton production generally will have a direct influence upon the supply of cottonseed.

Figure 3 presents cotton production, use, and carry-over data

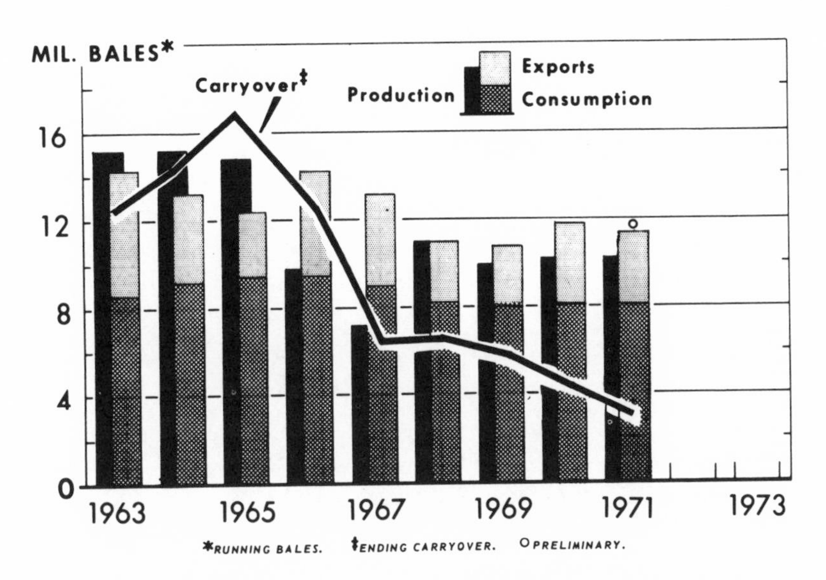

FIGURE 3 Cotton production, use, and carry-over.

from 1963 through 1971; it will be apparent that cotton production sharply decreased in 1966 and 1967 due to changes in government programs. This decrease in production was ultimately responsible for a decrease in cotton in storage. Cotton production in 1972 is estimated at approximately 13 million bales (Statistical Reporting Service, 1972).

Obviously, cotton production is controlled by textile mill consumption of fibers. Figure 4 gives data on fiber consumption over a 35-yr period. Growth of man-made fibers is evident; however, these data show that the growth trend seems to have leveled off. It is anticipated by many that future cotton production will continue at present production levels. Therefore, approximately 2 million tons of cottonseed meal/yr should be available as an alternative source of protein for animal production for many years to come.

Table 3 compares current utilization data for cottonseed meal and soybean meal (E. F. Hodges, personal communication). The principal markets for cottonseed meal are dairy and beef, whereas soybean meal is used widely in poultry and swine rations. Indeed, approximately one third of the cottonseed meal produced is successfully utilized in nonruminant rations.

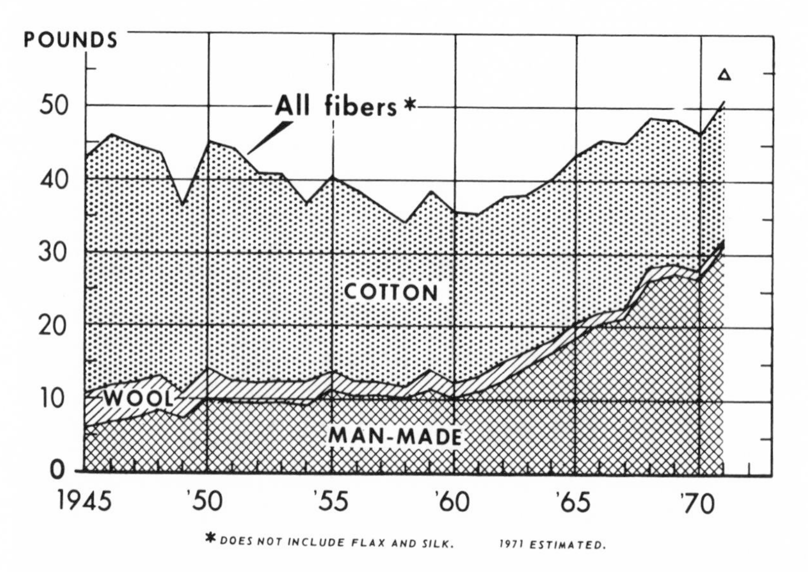

FIGURE 4 Mill consumption of fibers, per capita.

TABLE 3 Current Utilization[a,b]

Source	Soybean Meal (%)	Cottonseed Meal (%)
Beef	16.2	33.1
Dairy	10.4	19.2
Poultry	41.3	25.5
Sheep	1.6	3.1
Swine	18.6	9.7
Other	11.9	9.4

[a] From E. L. Hodges (unpublished data, 1972).
[b] Mean data for 1968–1970.

Still greater usage of cottonseed meal in rations of swine and poultry is possible. Shifts in meal utilization could significantly increase the tonnage of high-protein ingredients available to the nonruminant market.

Several factors influence the amount of cottonseed meal used in nonruminant rations; the principal one being gossypol. It has long been recognized that gossypol limits the usage of cottonseed meal in nonruminant rations. Extensive research, supported by practical experience, has shown that cottonseed meal may be a major oilseed supplemental protein source in nonruminant rations when care is taken to accept and adjust for limitations imposed by gossypol.

Research has defined the limits of gossypol permissible in rations supporting efficient performance. Knowing the maximum acceptable gossypol levels in a specific ration and the characteristics of the cottonseed meal to be included, one can formulate a ration to make maximum use of cottonseed meal. Cottonseed meal usage in most rations for nonruminants is influenced more by fiber, energy, and amino acid levels of the meal than by gossypol levels (Smith, 1970b).

By eliminating the prejudice against using cottonseed meal in nonruminant rations and including cottonseed meal as an alternative ingredient in the final ration, the level of cottonseed meal in the ration can then be based on the meal's ability to supply nutrients more economically than other feedstuffs and meet ration restrictions.

Certain nutrients vary depending on cottonseed meal production processes used (Smith, 1970a). These variations are important in determining the level of usage in scientifically formulated rations (Table 4). Screw-pressed cottonseed meals are relatively high in residual lipids and free gossypol and moderate to high in protein quality. Direct solvent meals are high in protein quality, moderate

TABLE 4 Composition of Cottonseed Meal by Process[a]

	Prepress Solvent (%)	Screw Press (%)	Direct Solvent (%)
Dry matter	89.9	91.4	90.4
Ash	6.4	6.2	6.4
Crude fiber	13.6	13.5	12.4
Ether extract	0.58	3.72	1.51
Crude protein	41.4	41.0	41.4
Gossypol, free	0.05	0.04	0.30
Gossypol, total	1.13	1.02	1.04
N-solubility	54.4	36.8	69.4

[a]From Smith (1970a).

in residual oil, and are highest in free gossypol, although there are meals that do not fit these general observations. It should be pointed out that protein quality, as used here, refers only to the availability of essential amino acids for poultry and swine.

The need to re-evaluate cottonseed meal as an alternative protein source is evident when changes in methods of processing are considered. Approximately 95 percent of domestic cottonseed processing in the 1940's was by hydraulic pressing. In 1965, 51 percent of the seed was processed by screw presses, 25 percent by prepress solvent extraction, 22 percent by direct solvent extraction, and only 2 percent by hydraulic pressing. The trend toward solvent extraction is continuing at the expense of screw-pressing. Estimates for the 1971 crushing season indicate that approximately 40 percent of the total cottonseed crushed was in screw-press mills and 30 percent each in direct and prepress solvent mills. These changes in processes, together with the compositional variation between processes and increased understanding of gossypol, emphasize the need to continually re-evaluate cottonseed meal as an alternative protein source.

If cottonseed meal is to increase available protein supplies for animal production, it must replace some of the soybean meal in rations. It is interesting to compare the two oilseed meals irrespective of process. Cottonseed meal is normally merchandized with less protein and more lipid and fiber than is contained in soybean meal. Cottonseed protein contains less of the amino acids, lysine, isoleucine, leucine, and, possibly, threonine and valine than soybean meal; it equals or exceeds soybean meal in all other critical amino acids. These differences in nutrients are directly involved in determining the competitive price and use levels for these two oilseed proteins in nonruminant rations.

To determine what restrictions should be placed on free gossypol levels in the various rations, an extensive review of the literature was undertaken. From the review, it has been possible to arrive at certain sound generalizations concerning feeding operations. These are summarized below.

Ruminants

In animals with a functioning rumen, dietary gossypol has not been shown to affect performance. The rumen fermentation action and by-products render the gossypol inert. Symptoms of gossypol toxicity, however, have been demonstrated whenever gossypol has been injected into the bloodstream.

Broilers

Tolerance to gossypol may be affected by age and strain of birds, level of dietary protein, iron salts, alkaline materials, and possibly other ration components. These variables account for the wide variation in gossypol tolerances reported in the literature. Broiler performance is not affected by dietary free gossypol levels up to 150 ppm (0.015 percent). Levels up to approximately 400 ppm (0.04 percent) in the ration may be fed successfully if ferrous sulfate is added at a 1:1 iron-to-free gossypol weight ratio.

In most cereal-based broiler rations, lysine content is critical. Attention has to be given to maintaining adequate lysine levels when cottonseed meal is included in the ration. A simple corn–soybean meal broiler ration is marginal in lysine. Since cottonseed meal contains approximately two thirds of the lysine as soybean meal, the importance of this key ingredient is readily apparent. Combinations of cottonseed meal and soybean meal in a nutritionally adequate ration often have been shown to support greater performance than either when fed alone.

Layers

Nutrient levels in cottonseed meal are compatible with requirements of the laying hen. Egg production is unaffected by dietary free gossypol up to 200 ppm (0.02 percent). Dietary levels of free gossypol up to 50 ppm may be fed without egg yolk discoloration. When higher levels of free gossypol (up to 150 ppm) are fed, protection against yolk discoloration is provided by supplementing with iron at a 4:1 weight ratio to gossypol. Since cottonseed lipids have been shown to enhance yolk discoloration by gossypol, it is advisable to minimize residual lipid levels in the ration.

Swine

Performance of growing–finishing swine is not affected by feeding rations containing up to 100 ppm (0.01 percent) free gossypol. A 1:1 weight ratio of iron to free gossypol may be used to inactivate free gossypol in excess of 100 ppm. Maximum level of supplemental iron recommended is 400 ppm.

It is usually safe to conclude that the low free gossypol level in screw-press and prepress solvent meals is not a significant factor in determining use in practical swine and poultry rations. The upper cottonseed meal level is usually determined by the lysine–fiber–energy variables, assuming that the meal is economically competitive.

The higher free gossypol levels in direct solvent meals may limit quantities in selected rations or may necessitate the need to include iron salts to inactivate the free gossypol present. Even with the gossypol restrictions imposed on high gossypol meals, however, sizable quantities of cottonseed meal may be used safely in nonruminant rations.

In summary, cottonseed meal may contribute substantially to needed protein supplies, if gossypol prejudices are replaced with the technical information now available.

OTHER OILSEED MEALS

There are several other oilseed meals that contribute to the protein supplies. They include peanut, linseed, safflower, sesame, sunflower, copra, castor, and rapeseed meals. While production volume of these meals is limited, they do provide a significant tonnage of high-protein ingredient.

There is considerable information in the literature on the utilization of these oilseed protein meals. Like other ingredients, they are evaluated in computer-formulated rations on the basis of cost/unit nutrient. Like cottonseed meal, most of these oilseed meals contain physiologically active and naturally occurring compounds that affect animal production if the compound is not inactivated or its intake limited. These limitations or ration restrictions have been adequately researched. When the computer is accurately programmed, rations may be formulated from these oilseed meals that will support acceptable performance.

The future prospects for increased production of these oilseed meals are directly related to vegetable oil markets. With healthy markets for vegetable oils, production of the oilseed meals is probable. It should be stressed that production of these oilseed meals is dependent upon oil movement or government programs to encourage the oilseed production. If production is encouraged, these oilseed proteins may contribute significantly in satisfying animal protein needs.

CONCLUSION

Oilseed proteins are the major source of supplemental protein in livestock rations. Oilseed protein supplies available for livestock production may be expanded by increasing production of the oilseed or by changing utilization patterns. The production of oilseeds is not controlled by the need for protein, but rather how these oilseeds fit into complex interrelationships of world trade–product markets and government programs. Therefore, the factors controlling increased production of the various oilseeds are indeed complex. Changes in oilseed utilization patterns offer the best immediate solution to supplying protein to critical areas of need. Increased amounts of protein may be made available to nonruminants through greater usage of unconventional nitrogen sources in ruminant rations that are less critical in their protein needs.

REFERENCES

Economic Research Service. 1972. Feed situation. Report FdS-244. Economic and Statistical Analysis Division, ERS, USDA, Washington, D.C. 27 p.

Smith, K. J. 1970a. Nutrient composition of cottonseed meal. Feedstuffs 42(16):19–20.

Smith, K. J. 1970b. Practical significance of gossypol in feed formulations. J. Am. Oil Chem. Soc. 47:448–450.

Statistical Reporting Service. 1972. Crop production. Report CrPr 2-2 (8-72). Crop Reporting Board, USDA, Washington, D.C. 17 p.

U.S. Department of Agriculture. 1971. 1971 Handbook of agricultural charts. Agricultural handbook #423. USDA, Washington, D.C. 146 p.

III
Identification and Evaluation of New Sources of Protein

J. H. Maner

INVESTIGATION OF PLANTS NOT CURRENTLY USED AS MAJOR PROTEIN SOURCES

It is well known that over one half of the world's human population go to bed hungry each night and that more people of the world today are suffering from malnutrition than are adequately nourished. Not only are too few calories available to this portion of the population, but the diet consumed is grossly deficient in protein and the essential amino acids required for normal growth, health, and reproduction. A similar, if less emotional situation, also exists within the world's animal population. Because of incomplete and inadequate animal diets, large quantities of both energy and protein are inefficiently utilized, or, more specifically, wasted. This inefficiency and waste can be well illustrated with data obtained from a small farm study on the north coast of Colombia. Traditional feeding methods utilize normal corn fed on the ground to pigs that are allowed to roam freely throughout the village. Under this feeding and management system, 9.6 kg of corn containing 0.96 kg of corn protein (9.6 kg × 10 percent crude protein) are required to produce a kilogram of live weight gain. On the other hand, utilization of an adequate diet properly fortified with protein, vitamins, and minerals reduced the feed requirement per kilogram of gain to only 2.6 kg of diet containing only 0.42 kg of protein. The point is that, for lack of a small quantity of quality protein, much of the available protein of lower quality is literally being poured onto the ground in the form of urinary nitrogen, i.e., the by-product of protein and amino acid catabolism. This same phenomenon is even more vividly demonstrated by members of the human population who suffer from the protein deficiency disease, kwashiorkor.

There is a growing awareness that the protein deficit problem is one of the most critical and difficult aspects of the total food and feed problem. There is no single solution to this complex world crisis. Every avenue of approach must be utilized in the hope that one or more of the approaches will provide the breakthrough necessary for staving off starvation and famine of an ever-increasing world population.

A survey of plants that have the protein content necessary to place them into a special category as a protein source provides only a limited number of prospects. These include the oilseeds, the grain legumes, the forage legumes, and various others such as copra, palm kernels, tung nut, and the castor bean. In this paper, only a few of the less-utilized oilseeds and the most important grain legumes will be discussed. Although the forage legumes and other sources of leaf protein offer great potential, this will be discussed elsewhere in this symposium.

OILSEEDS

Oilseeds, including the oil-bearing tree fruits, supply more than one half of the world's fats and oils and all but a small portion of its protein feed. Soybeans, cottonseed, and peanuts are the leading oilseeds and represent more than 80 percent of the world's supply of oilseed protein (Table 1). Sunflower, sesame, rapeseed, and copra represent less important sources of protein and have received less attention until recently. Investigations of some of these oilseeds justify a summary of their value as protein sources.

Sunflower Meal

Sunflower meal is produced from the seed of the sunflower plant (*Helianthus annuus*). The Soviet Union produces more than 6.6 million metric tons or some 66 percent of the total world production. Europe produces 1.58 million metric tons and Argentina 1.1 million metric tons (FAO, 1970).

TABLE 1 World Production of Major Oilseeds[a]

1970	Million Metric Tons
Soybeans	46.520
Cottonseed	22.060
Peanuts	18.140
Sunflower seed	9.650
Rapeseed	6.500
Linseed	4.140
Sesame seed	1.860

[a]From FAO (1970).

The energy content and protein concentration of sunflower meal varies according to the composition of the seed, the quantity of hulls present, and the method of processing (Smith, 1968). There has been a marked improvement in meal quality during the past few years due to advances in oil extraction and meal processing that has lowered the crude fiber and ether extract content of the finished product. The higher protein meals are produced by removing the largest quantity of hulls possible before processing and screening of the meal after extraction. The average composition of both expeller and solvent-extracted meals is presented in Table 2 and an amino acid profile is given in Table 3. Expeller-processed meals contain more fat and fiber and lower quantities of crude protein than do meals produced by solvent extraction. Solvent extraction and low-temperature processing improves the quality of the protein by reducing the destruction or loss of lysine, methionine, threonine, and tryptophan (Renner *et al.*, 1953). Dry, high-temperature processing causes a marked reduction in lysine content and availability. Arginine and tryptophan are also reduced by high-temperature processing. Soviet work (Tkacev *et al.*, 1964, 1965), reported by Smith (1968), indicates that, when meals were processed at three different temperatures (severe, 124–130 °C; normal, 114–124 °C; and mild, 105–114 °C), the mild heat-treated meals produced the highest digestibility of organic matter, ether extract and nitrogen-free extract, highest average daily gains, and lowest feed per unit gain when fed to swine.

The energy content and protein concentration of sunflower meal varies with the quantity of hulls present. The higher protein meals are produced by removing the largest quantity of hulls possible before processing and the screening of the meal after extraction. New vari-

TABLE 2 Chemical Composition of Some Oilseed Meals

Material Analyzed	Moisture (%)	Protein (%)	Ether Extract (%)	Fiber (%)	Ash (%)	Nitrogen-Free Extract (%)	Reference
Coconut meal, expeller	10.1	20.9	5.8	10.5	6.5	46.2	Creswell and Brooks (1971a)
Rapeseed meal, expeller	6.0	35.2	7.0	15.5	6.8	29.5	Wetter (1965)
Rapeseed meal, solvent	8.0	40.5	1.1	9.3	7.2	33.9	Wetter (1965)
Safflower meal, solvent	6.8	19.8	0.3	40.1	4.5	28.5	Shimada and Brambila (1966)
Sesame meal, hydrolic	7.8	38.7	10.8	6.4	9.7	26.9	Zaghi and Bressani (1969)
Sesame meal, solvent	3.8	45.1	0.7	5.2	13.2	32.0	Zaghi and Bressani (1969)
Sunflower meal, expeller	7.0	41.0	7.6	13.0	6.8	24.6	Smith (1968)
Sunflower meal, solvent	7.0	46.8	2.9	11.0	7.7	24.6	Smith (1968)

TABLE 3 Amino Acid Content of Some Oilseed Meals[a]

Item	Arginine	Histidine	Isoleucine	Leucine	Lysine	Methionine	Cystine	Phenylalanine	Threonine	Tryptophan	Valine	Reference
Coconut meal, expeller	9.37	1.96	2.87	5.78	2.29	1.77	1.14	3.87	3.15	–	4.25	Creswell and Brooks (1971a)
Rapeseed meal, expeller	5.09	2.04	3.71	6.45	4.39	1.88	–	3.74	4.08	0.94	4.76	Wetter (1965)
Rapeseed meal, solvent	5.50	2.69	3.65	6.72	5.39	1.93	–	3.82	4.24	1.23	4.84	Wetter (1965)
Sesame meal, expeller	11.91	2.21	4.27	6.92	2.76	2.65	–	4.73	3.64	1.91	5.06	Catron and Hays (1958)
Sunflower meal, expeller[b]	9.4	2.1	4.0	6.1	3.3	1.6	–	4.2	3.2	1.0	4.8	Smith (1968)
Sunflower meal, expeller[b]	8.7	2.1	3.9	5.9	2.8	1.5	–	4.3	3.2	1.0	4.9	Smith (1968)
Sunflower meal, solvent	8.2	1.7	5.2	6.2	3.8	3.4	–	5.7	4.0	1.3	5.2	Smith (1968)

[a] Low-temperature processing.
[b] High-temperature processing.

eties of sunflower seed with lower percentages of hulls are now available.

Sunflower meal used to supply all of the supplemental protein in growing–finishing swine rations is unsatisfactory. Because of its very low-lysine content, sunflower meal produces gain and efficiency of feed utilization inferior to that produced by soybean meal, fish meal, peanut meal, or combinations of these ingredients (Pearson *et al.*, 1954; Delic *et al.*, 1963, 1964). The most efficient utilization of the meal is obtained by combining it with such supplements as fish meal or soybean meal, which are rich in available lysine, and by limiting its use principally to diets for finishing pigs or adult animals used for reproduction. Further studies are needed to determine the effect of supplementing sunflower meal diets with crystalline lysine.

Rapeseed Meal

Rapeseed meal is a by-product of the production of oil from rapeseed (also called rape, colza, or raps). A member of the Cruciferae or mustard family, many species of the genus *Brassica* are known; however, two of these species, *B. napus* L. and *B. campestris* L., are most commonly grown. Because of its adaptation to the temperate and subtropical zones, rapeseed is produced as a winter or cool season crop in Europe, Asia, and in the western hemisphere (Downey, 1965). Rapeseed is an export crop for Canada and a source of oil and animal feed for Chile. Two thirds of the world annual production of 6.5 million metric tons are produced in Asia, and 1.6 million metric tons are produced in Canada (FAO, 1970).

Rapeseed processed by modern methods yields approximately 40 percent oil and 50 percent oil meal or cake (Wetter, 1965). Crude protein levels of rapeseed meals range between 32 and 44 percent, the majority containing 35–37 percent. A typical solvent-extracted meal will contain 40.5 percent protein, 1.1 percent ether extract, and 9.3 percent fiber. Examination of the essential amino acid composition of modern extracted rapeseed meal indicates that the protein is suitable for livestock feeding. The balance and content of amino acids is superior to that of many other plant protein meals and compares favorably with soybean meal. Processing method can greatly influence the availability of amino acids and thus the nutritive value of the meal. The high heat of processing by the expeller method is associated with poor amino acid availability; however, a combination expeller and solvent extraction that is conducted at a lower temperature produces a higher quality meal (Young, 1965).

Although the amino acid composition would indicate that rapeseed meal can be utilized for swine, it contains undesirable substances that limit its usefulness as a major protein supplement for swine and that may exert considerable effect on its nutritional value (Bell and Belzile, 1965; Belzile *et al.*, 1963). These undesirable substances are related to its thioglucoside content, which upon enzymatic hydrolysis by the thioglucosidase, myrosinase, yields thiocyanates and oxazolidinethione, commonly referred to as mustard oils. These two compounds, and especially the oxazolidinethione, have been shown to exert a goitergenic effect on nonruminant animals, causing histological changes of the thyroid that can be only partially overcome by feeding iodinated casein. Low levels of these toxic substances consumed by pigs in diets containing rapeseed meal cause thyroid hypertrophy and increases the cellular components. Higher levels (supplied by 10–12 percent meal) cause a marked increase in both cellular components and glandular hypertrophy. High levels of rapeseed meal (more than 5 percent of the diet) may reduce palatability and consumption and therefore have adverse effects on the growth rate of the pig. When growth depression occurs as a result of high levels of rapeseed meal, it does not affect feed utilization and cannot be corrected by lysine supplementation but usually is associated with reduced feed intake (Hussar and Bowland, 1959; Bowland, 1965).

Results similar to those obtained with young pigs fed rapeseed meal have been observed with growing–finishing pigs (Manns and Bowland, 1963; Bowland, 1965), although to a lesser degree. Replacement of 25 percent of the protein supplement with rapeseed meal does not significantly affect feed intake, rate of gain, or feed efficiency, but when higher levels are utilized both rate of gain and feed efficiency are depressed (Devilat, 1965; Esnaola and Ochoa, 1970).

Gilts fed high levels (10–12 percent) of rapeseed meal have been shown to require more estrous cycles for conception, to farrow lighter and smaller litters, and to wean fewer and lighter weight pigs than similar animals fed diets not containing this meal. These reports (Bowland, 1965; Bowland and Standish, 1966) indicated that not only is sexual maturity delayed but also lactation is depressed when rapeseed meal represents a major portion of the protein supplement of the diet.

Because of the poor reproductive performance observed when gilts are fed diets containing one half or more of their protein supplement as rapeseed meal, extreme caution should be taken when including this meal in diets for breeding animals. It is recommended that not more than 3 percent solvent-extracted rapeseed meal should be used in the diet of breeding swine during pregestation, gestation, and lactation (Bowland, 1965).

Expeller-processed rapeseed meals have reduced growth rate and feed efficiency in both chicks and poults. Thyroid enlargement was noted even when as little as 5 percent rapeseed meal was included in the chick diet (Clandinin, 1965). However, low-temperature, expeller-processed rapeseed meal has been shown to approach or equal soybean meal for chick growth promotion. Where slow growth has been observed, it has been the result of feeding overheated meal. Prepress-solvent or solvent-processed rapeseed meal not subjected to excessive heat has been found equivalent to soybean meal for chicks and can be added to chick rations at a rate of 10–15 percent of the total diet.

Production, feed conversion, fertility, and hatchability have been shown to be satisfactory, even when up to 10 percent prepress-solvent or solvent-processed rapeseed meal has been included in the diet.

The results of experiments with cattle and sheep indicate that rapeseed meal can be fed in practical rations with no serious difficulty. Ruminant animals do not develop enlarged thyroid glands and no adverse effects on the rate of gain or reproduction have been noted when solvent-extracted rapeseed meal was fed at high levels (Robblee, 1965). Neither yield nor milk flavor is affected by inclusion of rapeseed meal in dairy rations.

Sesame Meal

Sesame, known botanically as *Sesamum indicum* (also called benne, til, gingerly, simsin, gergelin, and in Latin America, ajonjolí), is probably one of the oldest of the cultivated oilseed crops (Cornelius and Raymond, 1964). World production of sesame amounts to 1.86 million tons (FAO, 1970), with China and India, the chief producers, accounting for about 47.5 percent of the total. Sesame also grows in many other countries in the Far and Near East and on the American continents. Sudan is the leading African producer and Mexico the leading American producer. Most of the countries where sesame is grown consume the bulk of the crop domestically; international trade in seed and oil is small compared with the quantities produced.

The seeds—which vary from white through black in color, according to variety—are small, usually 2–3.5 g/1,000 seeds and are contained in capsules that tend to split open and shed when ripe (Johnson and Raymond, 1964). Because of this tendency to pop open, practically all the harvesting is carried out by hand. Nonshattering lines and varieties must be developed if mechanical harvesting is to be successfully developed and thereby allow the development of sesame as a competitive crop in areas of high labor cost.

The seed has a thin hull that can be separated from the kernel in decorticating machines or by soaking and rubbing. Normally, the seed is milled without decortication to remove the oil (44–54 percent present in the air-dried seed). The sesame meal produced from the whole seed contains both the kernel and the thin hull.

Sesame meals processed by the screw press or hydraulic methods contain higher levels of fat and lower levels of protein than those produced by solvent extraction (Mitra and Misro, 1967) (Table 2). Chemically, those processed by pressure contain 7–8 percent moisture, 38 percent crude protein, and 10–11 percent oil, whereas those processed by solvent extraction contain less than 1 percent oil and over 45 percent crude protein (Zaghi and Bressani, 1969). Sesame meal is an excellent source of methionine and tryptophan but contains only about 2.7 percent lysine, which does not vary greatly from normal corn (Table 3).

As a protein supplement for swine diets based on cereal grains, sesame meal should be blended with such other supplemented protein sources as soybean meal or fish meal (Maner and Gallo, 1963, 1970; Gallo and Maner, 1970). Sesame meal can replace one half, or all, of the soybean meal in corn–soybean diets containing 4 percent fish meal and can be substituted for up to 10 percent of the soybean meal in a 16 percent protein corn–soybean meal diets for growing–finishing swine (Gallo and Maner, 1970; Maner and Gallo, 1970). Because of the low levels of lysine in both, cottonseed and sesame meals or blends of these two supplements are inadequate for supplementing cereal-based growing–finishing diets. However, 50–50 blends of these two protein sources fed along with cereal grains and 5–6 percent fish meal produce satisfactory gains and efficiency of feed utilization (Hervas *et al.*, 1965).

Coconut Meal

Coconut meal or copra meal, the residual material from extraction of oil from the dried meat (copra) of the coconut (*Coco nucifera* L.) is available in many parts of the world. The Philippines, Indonesia, India, Ceylon, and Malaya produce the bulk of the world supply. Coconut meal in some areas may be one of the few protein sources available for livestock feeding. Even though the meal contains only moderate levels of crude protein, it is an economically important source of protein in areas where other commonly used sources are not available.

Coconut meal contains between 20 and 26 percent crude protein, about 6 percent of ether extract when expeller processed, and about 10 percent of crude fiber (Table 2). The digestibility of the various constituents of coconut meal is high, except for the protein fraction. Expeller-extracted coconut meal has 83.7 percent digestibility of the dry matter; ether extract, 100 percent; nitrogen-free extract, 94.1 percent; and energy, 85.4 percent (Creswell and Brooks, 1971). Reported values for protein digestibility differ widely. Hawaiian workers (Creswell and Brooks, 1971) reported a protein digestibility of only 50.7 percent with pigs, while Philippine work (Loosli *et al.*, 1954) has shown the crude protein fraction to be 73 percent digestible. The variation in protein digestibility may be due to differences in processing methods and more specifically to processing temperature. Butterworth and Fox (1963) heated cold-solvent-extracted coconut meal for 30 min at oven temperatures from 40 to 150 °C and clearly demonstrated that each increase in treatment temperature caused a corresponding linear reduction in protein digestibility when the final material was fed to rats (Table 4). There was also a corresponding reduction in available lysine and net protein utilization as the temperature was increased. It has also been shown (Loosli *et al.*, 1954) that expeller processing reduces the digestibility of the protein. In these studies, fresh coconut meals had a crude protein digestibility of 83.8

TABLE 4 Effect of Various Degrees of Heat Treatment on the Composition and Nutritive Value of Coconut Meal[a]

Temperature of Treatment (°C)	Protein Content (%)	Available Lysine (g/16 g N)	Coefficient of True N Digestibility (%)	Net Protein Utilization
40	25.3	3.29	77.7	45.9
90	25.8	3.09	78.3	41.0
105	26.3	2.81	74.6	36.1
120	25.6	2.34	73.3	35.8
135	26.4	1.63	68.2	33.9
150	26.3	1.12	56.1	17.1

[a]From Butterworth and Fox (1963).

TABLE 5 Effect of Level of Expeller-Processed Coconut Meal on the Performance of Pigs[a]

Level of Coconut Meal (%)	Average Daily Gain[b] (kg)	Average Daily Feed (kg)	Feed/Gain (kg)	Loin Eye Area[b] (cm^2)
0	0.76	1.88	2.51	27.1
10	0.74	1.97	2.68	26.9
20	0.65	1.80	2.75	25.5
40	0.46	1.49	3.27	22.3

[a]From Creswell and Brooks (1971b).
[b]Values are not significantly different ($P < 0.05$).

percent and expeller-processed meal protein was only 73.4 percent digestible.

Different levels of coconut meal have been used (Creswell and Brooks, 1971; Grieve *et al.*, 1966) to supply a portion of the protein in complete mixed diets for growing–finishing pigs. Each increase in level of coconut meal used to replace soybean meal caused a decrease in gains and an increase in feed required per unit gain. Dietary levels up to 20 percent caused only minor reductions in pig performance (Creswell and Brooks, 1971); however, 40 percent dietary coconut meal caused a 39.5 percent reduction in gains and a 30 percent increase in feed required per unit gain (Table 5). Loin eye measurements were also reduced at the highest level of substitution. Although it would appear that lysine might be deficient in the rations prepared with expeller-processed meal, supplementary lysine was without effect when added to diets containing 20 or 40 percent coconut meal (Creswell and Brooks, 1971), and, as suggested by these same authors, some factor other than lack of adequate protein and lysine may be responsible for the growth depressing effect of coconut meal.

Safflower Meal

Safflower (*Carthamus tinctorius*) has long been cultivated as an oilseed crop. Although the seeds contain 36–40 percent oil, the hulls constitute about 40 percent of the seed. After the oil is removed, the product that remains contains only 18–22 percent protein and about 60 percent hulls. The hulls contain more than 20 percent lignin, which greatly limits the energy concentration and utilization of the meal by swine and other classes of livestock.

Attempts have been made to improve the quality of safflower meal by the physical separation of the hulls from the kernel and through the development of thin hull varieties through plant breeding. Although the removal of the hull results in a seed with 60–70 percent oil and a seed cake of about 60 percent protein, in practice it is very difficult to effect a complete separation because of the hardness of the seed coat and the extreme softness of the kernel. Meals produced from regular varieties with partial hull removal have resulted in feeds with over 40 percent protein and about 15 percent fiber.

The protein of safflower oil meal is a very poor source of lysine and is somewhat deficient in the sulfur-containing amino acids, methionine, and cystine (Kratze and Williams, 1951).

Although definitive data are not available to indicate the true nutritive value of safflower meal for swine, Mexican workers (Shimada and Aguilera, 1966; Shimada and Brambila, 1966; Bravo and Cabello, 1969) have conducted a limited number of trials that tend to characterize the meal. Safflower meal, containing 30.9 percent fiber and 18.8 percent crude protein, was substituted for soybean meal in sorghum-based diets for growing pigs (17.4–56.5 kg) (Shimada and Aguilera, 1966). Safflower meal supplied 0, 12.5, and 25 percent of the total supplemental protein, substituting for both soybean meal and sorghum. Twenty-five percent safflower meal significantly reduced both gains and efficiency of feed utilization. At the 12.5 perdent level of substitution, however, no changes in pig performance were observed.

In subsequent studies Shimada and Brambila (1966), employed a 19.8 percent protein safflower meal containing 40.1 percent crude fiber. In diets containing fish meal (25 percent of the supplemental protein) and cottonseed meal (50 percent of the supplemental protein), the addition of safflower meal (25 percent of the protein) reduced gains and increased feed per unit of gain when compared with those produced by soybean meal. The authors suggest a maximum level of 7–8 percent safflower meal in the diet of growing pigs to supply not more than 12.5 percent of the supplemental protein.

Bravo and Cabello (1969) working at the same station with finishing pigs, used dietary levels of 15, 20, and 25 percent safflower meal in grain–molasses basal diets. Although gains were not greatly different among treatments, overall performance of all groups were poor and feed required per kilogram of gain increased with each increase of safflower meal. The authors suggest that not more than 10–12 percent of the diet for finishing pigs should be supplied by safflower meal.

Because of its low protein and high fiber, sesame meal is of limited value in growing–finishing swine rations. It would appear that it could best be utilized in rations for gestating sows whose requirements for energy are low.

GRAIN LEGUMES

In the search for solutions to the protein problem, the food and grain legumes have many advantages and offer a wide range of germ plasm of wide environmental adaptability with which to meet the protein needs of both the human and animal population. Of the 600 genera, with more than 13,000 species that comprise the family Leguminosae, only some 20 of the seed-producing species are utilized regularly in appreciable quantities (Aykroyd and Doughty, 1964). However, with the exception of soybean and peanuts, which contribute significantly to the total supply of feed protein, the majority are low yielding and contribute only in a limited way to the world's total protein supply (Table 6). Because of low yields, these crops cannot in many cases compete economically with other higher yielding crops such as maize, rice, wheat, barley, and cassava. The amount of of agronomic research conducted on the grain legumes has, in most cases, been limited. Production levels, therefore, remain low because of the inherent characteristics of the varieties grown, damage from diseases and insect pests, and poor cultural practices. The majority of the food and grain legumes, with the exception of soybeans (*Gly-*

TABLE 6 World Production of Certain Legumes–1970[a]

	Production (Million Metric Tons)	Yield (kg/ha)
Soybeans	46.521	1,330
Peanuts	18.144	950
Dry beans	11.682	500
Dry peas	10.209	1,120
Chickpeas	7.013	700
Dry broad beans	5.181	1,100
Pigeon peas	1.960	670
Cowpeas	1.180	380
Other pulses	3.165	550

[a]From FAO (1970).

cine max), peanuts (*Arachis hypogaea*), beans (*Phaseolus vulgaris*), and field or garden peas (*Pisum* sp.), which are important in the developed world, are of importance only within the developing countries. Those of major importance, except soybean and peanut, will be discussed.

Field Peas

Field peas (*Pisum satium* or *arvense*) also called peas, garden peas, and arvejas is an annual. Different varieties are grown for both human food and livestock feed. They thrive only in temperate regions or as cool season or hill country crops in the tropics and subtropics. Field peas are grown in many areas of the world but have been used extensively as a feed for livestock only in Russia and parts of Europe. The cracked and shriveled peas, not usable for human consumption, are now being used for livestock feed in the Pacific Northwest of the United States. These peas are highly palatable to livestock and especially to pigs of all ages. The peas, which contain 22–29 percent crude protein, are normally utilized to replace a portion of the protein supplement in swine rations. If the price of peas and cereals is similar, peas can be used as an economic grain substitute.

Peas were demonstrated to be inferior in protein quality to either meat meal or soybean meal when fed to growing–finishing pigs in dry-lot as the only protein supplement to cereal-based diets (Beeson and Hickman, 1945; Lehrer and Hodgson, 1946). Subsequent studies, however, have indicated that the growth depression and general poor performance of animals fed peas were most likely related to deficiencies of nutrient factors other than energy and amino acids.

Later reports (Cunha *et al.*, 1948; Warwick *et al.*, 1948) have shown the value of the field peas as both a protein and energy source for pigs of all classes. Research conducted in Australia (Lobban, 1963) indicate that satisfactory performance of growing–finishing swine can be obtained with 20–30 percent of the diet supplied by peas.

Kroening (1968) demonstrated that peas can supply all of the supplemental protein in cereal grain diets for growing–finishing pigs and that neither gains nor feed efficiency are improved by methionine supplementation as might be expected from the amino acid analysis of the peas used (Table 7). The level of total sulfur amino acids (methionine + cystine) is sufficient to meet the methionine needs of the growing–finishing pig if the level of peas does not exceed 60 percent

TABLE 7 Amino Acid Content[a] of Some Grain Legumes

Item	Arginine	Histidine	Isoleucine	Leucine	Lysine	Methionine	Cystine	Phenylalanine	Threonine	Tryptophan	Valine	Reference
Chickpeas (forage)	8.93	3.07	3.87	7.46	4.03	1.36	1.04	5.90	3.12	0.98	4.24	Abu-Shakra *et al.* (1970)
Cowpeas	–	–	3.00	4.70	4.10	1.00	0.80	3.30	2.50	0.60	3.50	Aykroyd and Doughty (1964)
Dry beans	–	–	3.60	5.40	4.60	0.60	0.60	3.50	2.70	0.60	3.80	Aykroyd and Doughty (1964)
Field beans	9.60	2.59	4.34	7.44	6.22	0.82	1.28	4.28	3.68	0.99	4.74	Henry (1970)
Field peas	7.15	2.62	3.59	6.28	6.52	0.76	0.55[b]	4.35	4.04	0.86[b]	3.87	Bell and Wilson (1970)
Pigeon peas	–	–	3.80	4.90	4.50	0.70	0.90	5.40	2.40	0.30	3.30	Aykroyd and Doughty (1964)

[a] Amino acids expressed as percent of protein.
[b] From Kroening (1968).

TABLE 8 Effect of Feeding Either Corn, Wheat, or Barley with Soybean Meal or Peas on the Performance of Growing Pigs[a]

Source	Corn	Wheat	Barley	Average
Average Daily Gain (g)				
Soybean meal	846	795	764	802
Peas	818	805	795	806
AVERAGE	823	800	780	
Feed/Gain				
Soybean meal	3.06	2.94	3.14	3.05
Peas	2.88	2.81	2.84	2.84
AVERAGE	2.97	2.88	2.99	

[a] From Kroening (1968).

of the 16 percent protein diet, based on corn, wheat, or barley (Table 8) (Kroening, 1968).

In Canadian studies (Bell and Wilson, 1970), field peas of cull grade replaced 0, 25, 50, 75, and 100 percent of the soybean meal–fish meal supplement of a swine-grower ration without reducing rate of gain, feed efficiency, or carcass quality. Methionine supplementation proved unnecessary when the pea protein replaced all of the soy–fish protein. This supports the findings of Moran *et al.* (1968), who found that for chicks the cystine level compensated for the deficiency of methionine in peas and that supplemented methionine was not needed. These same authors (Moran *et al.*, 1968) found that autoclaving (15 min, 121 °C) the whole pea substantially improved gain: feed ratio but had little effect on growth of 1- to 3-week-old chicks. Metabolizable energy, determined with adult roosters was increased by either autoclaving or high-temperature steam pelleting (90 °C) of the pea meal. Dry heating of the whole pea in an oven at 121 °C for 45 min caused a reduction in metabolizable energy.

For swine, it is not necessary to cook field peas as they can be fed raw with no ill effects (Kroening, 1968; Bell and Wilson, 1970). Raw peas can be satisfactorily utilized in swine diets with excellent results, even though Tannous and Ullah (1969) reported that field peas contain 80 units of hemagglutinizing activity and an antitrypsin activity of 8.4 units. Although autoclaving for 5 min at 121 °C reduces the antitrypsin activity to 2.25 and prevented rat mortality, the level present in raw peas does not appear to affect pig growth or efficiency of feed utilization. When dry heat was used, the performance of pigs fed diets containing 12 percent toasted seed was poorer than that of those fed the same level of untoasted peas (Thomke and Frölick, 1968). Cooked peas, used as a substitute for fish meal in free-choice supplement and corn rations, have also been shown to support pig performance less well than did the control ration (Rosa *et al.*, 1969).

Most of the genetic work on peas has been directed toward their yield, resistance to disease and insects, and freezing and canning qualities. The little genetic work that has been done on the nutritional value of peas has been limited to increasing their nitrogen content. Almost no work has been reported on the biological quality of legume proteins from different varieties of seeds. Bajaj and co-workers (1971) utilized rats to compare the protein quality of 28 different lines of field peas grown under similar field conditions. These different lines of peas, when given as a sole source of protein to weanling rats at 10 percent level of protein in an otherwise adequate diet,

TABLE 9 Chemical Composition of Some Grain Legumes

Item	Moisture (%)	Protein (%)	Ether Extract (%)	Fiber (%)	Ash (%)	Nitrogen-Free Extract (%)	References
Chickpeas (forage)	11.8	17.3	3.5	8.4	2.8	56.2	Shimada and Brambila (1967a)
Cowpeas	11.0	23.4	1.8	4.3	3.5	56.0	Aykroyd and Doughty (1964)
Dry beans	11.0	23.9	1.3	4.2	3.4	56.2	Aykroyd and Doughty (1964)
Field beans	11.0	23.4	2.0	7.8	3.4	52.4	Aykroyd and Doughty (1964)
Field peas	10.1	25.9	1.5	6.0	2.6	53.9	Kroening (1968)
Pigeon peas	11.0	20.9	1.7	8.0	3.5	54.9	Aykroyd and Doughty (1964)

varied from 18 to 78 percent of that of casein in ability to support growth and nitrogen retention. There was no correlation between protein quality and the protein content of the different pea lines as measured by nitrogen incorporation efficiency (NIE) or protein efficiency ratio (PER).

Chickpea

The chickpea (*Cicer arietinum* L.) is an annual herb with drought-resistant properties grown throughout the subtropics and in the cool season in the dry tropics. The chickpea (also called garbanzo or bengal gram) consists of two main types: a large-seeded variety used for human food in southwestern United States, Mexico, India, Pakistan, and several European countries and a small-seeded variety known as forage chickpea that is used in the central valleys of Mexico mostly for feeding swine. The protein and fat content of the large-seeded varieties are about 20 percent higher than those of the small-seeded varieties, and the small seeds contain about three times more fiber than the large seeds (Table 9).

Very little information is available on yield potential of the chickpea; limited trials in Iran indicate that yields in excess of 3,000 kg/ha can be obtained utilizing selected local varieties in experimental plots with irrigation and good crop management (Roberts, 1970).

The chickpea contains less protein than the soybean and certain other pulses, but it can be used for swine feeding without cooking or heating (Shimada and Brambila, 1967b). The chickpea is a fairly good source of lysine but a relatively poor source of tryptophan, methionine and cystine, containing only 0.5 percent of the protein as tryptophan and 1.7 percent of the sulfur amino acids (Table 7).

Although it has been reported (Brambila and Shimada, 1967) that autoclaving improves the utilization of chickpeas by the chick, peas autoclaved at 115 °C for 30 or 60 min reduced the dry matter, energy, and nitrogen digestibility when fed to pigs (Shimada and Brambila, 1967b). Although autoclaving had no effect on performance of pigs receiving diets containing 88.8 percent chickpeas, Tannous and Ullah (1969) have shown that the raw chickpeas show no hemagglutinating activity but have a high antitrypsin activity, which is reduced–but not completely eliminated–when the peas were autoclaved for 20 min at 121 °C. Mexican workers (Shimada and Brambila, 1967a,b,c) in a series of three experiments have studied the chickpea as a source of both protein and energy for the growing pig. Levels of 0–88.8 percent chickpeas have been used to replace a mixture of corn and soybean meal in 16 percent protein diets or to replace portions of a sorghum–cottonseed meal–fish meal diet. At all levels of substitution, pigs performed equally well as those fed control diets. The raw chickpeas supplied adequate protein, energy, and amino acids to support satisfactory growth and efficiency of feed utilization, even when pea meal supplied the major portion (88.8 percent) of the diet. However, supplementation of the diets with 0.2 percent methionine improved both growth and feed conversion in the 88.8 percent chickpea diet. Higher levels of methionine supplementation tended to decrease gains (Shimada and Brambila, 1967a). Although not tested in these studies, it would appear that, at the higher levels of methionine supplementation, tryptophan may become the limiting amino acid (Pennacchiotti and Schmidt-Hebbel, 1966).

Field Bean

The field bean (*Vicia faba* L.), also known as horse bean or broad bean, is an annual herb that is grown in temperate and subtropical zones throughout the world mainly as a source of dry beans. Field beans are difficult to grow in the tropics. The relatively high protein content (24–31 percent) (Eden, 1968) and the general absence of inhibitory factors make the field bean a satisfactory feed for both growing–finishing or reproducing swine. Although it has not been used extensively as a pig feed, it is gaining popularity as a partial replacement for the supplemental protein for many classes of livestock.

The field bean, as are most legume seeds, is deficient in methionine (Table 7) but contains high levels of cystine that partially overcome the deficiency of methionine. It is an excellent source of lysine and a relatively satisfactory source of tryptophan. The chemical composition (Table 9) of the field bean indicates that it contains very little ether extract (1.5 percent) and only a moderate level of fiber (8–8.4 percent).

In diets for growing–finishing pigs, levels of up to 30 percent have been used successfully to replace dietary fish meal or soybean meal. Spanish workers (Balboa *et al.,* 1966) used 30 percent bean meal with and without 0.17 percent DL-methionine to replace 11 percent soybean meal in a corn–fish meal–soybean meal-type ration. The control, 30 percent bean meal and 30 percent bean meal plus methionine, diets contained a total methionine plus cystine level of 0.55, 0.43, and 0.60 percent, respectively. When these diets were fed *ad lib* in twice-daily feeding to growing pigs (25–60 kg), average daily gains and feed conversion ratios were depressed when bean meal replaced soybean meal unless the diet was also supplemented with methionine. When supplemented with 0.17 percent DL-methionine, performance was equal to that obtained with the control. No toxic effects were observed in any of the animals receiving bean meal. Luscombe (1969) (cited by Clarke, 1970) reported that performance of growing–finishing pigs (32–91 kg) were not affected by partial or complete replacement of the soybean meal with field bean meal in the diets. Neither growth nor feed efficiency were improved by the addition of methionine.

Although dry matter digestibility was 83–84 percent and digestibility of nitrogen was 83–87 percent when beans made up 20–40 percent of growing pig rations, Livingstone *et al.*, (1970) demonstrated a reduction in growth and an increase in feed per unit gain when field beans replaced one half or all of the fish meal in a barley–fish meal ration. Although, theoretically, 2.4 kg of beans should supply similar protein and the same quantity of lysine as 1 kg of fish meal, both gains and feed efficiency were significantly improved when a 3:1 bean to fish meal replacement was made as compared with a 2:1 replacement; however, even the higher replacement rate failed to support performance equal to that of the control, barley–fish meal ration.

Studies reported by Navratil (1965) clearly demonstrate that satisfactory performance can be obtained with pigs fed diets based on cereals, beets, and potatoes if bean meal replaces one half or all of the supplemental protein. Equal performance was obtained when bean meal replaced 50 percent of the supplemental protein from yeast and oil meals in diets for pigs up to 50 kg, and when bean meal supplied all of the supplemental protein from 50 to 90 kg liveweight. Bean meal additions of 18.5–25.0 percent to all-cereal diets improved gains of growing–finishing pigs by 65 percent and rate of feed conversion by 25 percent.

In practice, the level of bean meal substitution and need for methionine supplementation will depend upon the type of diet used and the source of supplemental protein.

Although digestive disorders have been previously indicated to occur when immature beans were fed to pigs, this does not appear to be a problem when dried, mature beans are fed either immediately after harvest or after extended periods of storage. Clarke (1970) fed newly harvested and dried mature beans to pregnant sows at an inclusion rate of 10 percent of the diet and observed no digestive upsets or adverse effect on the litters produced. Similarly, Luscombe (1969) observed no ill effects on the health of 23-kg pigs fed freshly harvested beans in the pods.

Beans

The genus *Phaseolus* covers a large number of species including *Phaseolus vulgaris* (common bean, dry bean, snap bean, kidney bean, haricot bean, navy bean), *P. aureus* (mung bean), *P. mungo* (urd bean), *P. lunatus* (lima bean), *P. calcaratus* (rice bean), and several others of minor importance. Of these, *P. vulgaris* is by far the most important in worldwide acreage, geographic distribution, and consumption (Roberts, 1970).

The common bean is native of Central America but is grown on all continents as a source of protein for the human population. Major areas of production are the north central and north western states of the United States, the medium to high elevations of Central and South America, Europe, and in countries of the Near and Middle East.

Although yield potential for the annual herbaceous plants are high (4,000 kg/ha), reported annual yields range from 300 to 1,500 kg/ha. The dry beans are susceptible to a wide range of insects and diseases, especially in the hot, humid tropics. The bacteria and fungi are more important in the lowland humid tropics and subtropics, while viruses are generally more severe in the drier climates (Roberts, 1970).

Although the primary activity in food and grain legume breeding research is to increase yields of grain so that the crop can become economically competitive, limited research is being directed to the improvement of protein content and quality (Rutger, 1970; Kelly, 1971; Leleji, 1971; Ries, 1971; Porter, 1972). Results to date indicate that yield is inversely correlated with percent protein, that the genotype of the maternal plant exercises control of the percent protein of the dry bean (Leleji, 1971; Porter, 1972), that protein quality as measured chemically by assaying for methionine or elemental sulfur is determined genetically, and that sufficient variation exists within the species to permit improvement through hybridization and selection (Kelly, 1971; Porter, 1972). Porter (1972) has shown that the nutritive value of some beans follows a trend similar to that of their elemental sulfur content (Tables 10 and 11). Silbernagel (1970) states that "years of extensive and expensive research will be required, but eventually we should be able to develop a bean that is as good a source of dietary protein as meat."

TABLE 10 Effect of Sulfur and Sulfur Amino Acids (S-AA) in Dry Beans on Protein Utilization[a]

Variety	Methionine (%)	Cystine (%)	Total S-AA (%)	Sulfur (% of Protein)	PER
Casein	–	–	–	–	2.81
Calima	1.21	1.15	2.36	0.91	1.96
Nicer 59	1.13	1.13	2.26	0.99	1.76
Cargamanta	1.26	0.75	2.01	0.71	1.36

[a]From Porter (1972).

As early as 1920, dry beans were shown to have a marked toxic effect on rats (Johns and Finks, 1920). Toxic factors identified in raw beans include a trypsin inhibitor (Bowman, 1944), a factor that inhibits protein synthesis in skeletal muscles (Kakade *et al.*, 1968), and a hemogglutinin (Rigas and Osgood, 1955). The trypsin inhibitor caused a significant pancreatic hypertrophy that was reduced by heat treatment of the raw beans (Bowman, 1944; Kakade and Evans, 1965, 1969; Conner *et al.*, 1971).

Kidney beans contain a medium to high level of hydrocyanic acid. Chilean studies (Jacob, 1967) indicate that four varieties of dry beans contained from 47.1 to 86.1 mg of hydrocyanic acid/kg of beans. Industrial cooking reduced the level but still did not completely eliminate the HCN from the beans. Consumption of these industrially processed beans caused digestive disturbances, which were not observed if the beans were first soaked overnight in water and boiled in new water for 2 hr.

TABLE 11 Effect of Crude Protein and Sulfur Content on Biological Value of Some Dry Beans[a]

Variety	Protein Content (%)	Sulfur Content (% of Protein)	Average Daily Gain (g)	PER
Casein	–	–	88.6	3.56
Husasanó	19.8	0.955	38.4	1.89
Calima	18.7	0.914	52.3	2.17
20295	18.6	0.914	42.3	1.88
Tui	20.0	0.860	36.5	1.99
Bunsi	20.8	0.832	59.7	2.31
Cargamanta	22.2	0.630	19.5	1.05
Mortiño	22.2	0.604	15.7	0.96

[a]From Porter (1972).

Nutritionally, dry beans, as most legumes, are deficient in the sulfur-containing amino acids and contain only low levels of tryptophan (Bandermer and Evans, 1963). Bressani (1969) has studied the variation in protein, methionine, cystine, and lysine content of selected lines of dry beans. Analysis of 268 samples of dry beans showed that the protein content varied from 16.8 to 28.2 percent. Analysis of 129 of these varieties for methionine, cystine, and lysine indicated a range of 0.087–0.355 percent methionine, 0.075–0.208 percent cystine, and 0.80–2.39 percent lysine. There was no correlation between total nitrogen content and the content of amino acids tested, which would support the suggestion that the protein quality of beans can be improved through selection and hybridization.

Purdom and Brown (1967) fed dry-bean diets containing 10 percent protein to growing rats for 28 days and reported that the unsupplemented bean protein supported growth and nitrogen retention that were inferior to those obtained with whole egg protein. Supplying the bean diet with 0.3 percent DL-methionine improved performance of the rats to a level not different from that obtained with the whole egg control; however, tryptophan supplementation was ineffective in improving performance in the presence or absence of methionine, even though amino acid analysis indicated that tryptophan was low.

Autoclaved beans fed to 2-day-old chicks at a level of 53 percent of the diet significantly reduced gains, efficiency of feed conversion, and total protein efficiency (g gain/g protein consumed). The addition of antibiotics (penicillin or Terramycin–HCl) to the bean diet significantly improved growth, feed efficiency, and total protein efficiency. A similar, but less marked effect, was also shown with methionine supplementation (Goatcher and McGinnis, 1972a,b), but no additive effect was noted when the bean diet was supplemented with both methionine and antibiotics. These same authors suggested that that beans have a depressing effect on growth (possibly by stimulating the growth of adverse microflora in the intestinal tract). Kakade and Evans (1964, 1966) obtained similar results with rats, showing that significant growth increases were obtained by the addition of antibiotic supplementation to diets containing dry beans. They attributed the improvement either to increased digestibility or absorption of nutrients, resulting from the antibiotics, or inhibiting of an enzyme that may be involved in liberating a bound growth inhibitor present in the dry beans.

Cowpea

Cowpeas (*Vigna sinensis* or *unguiculata*) is a herbaceous vine that probably originated in Africa but is widely cultivated in India, China, and Africa and, to a lesser extent, in both North and South America. Approximately 94 percent of the world production of 1.18 million metric tons comes from Africa. Although reported average world yields are extremely low, averaging only about 380 kg/ha, advanced cowpea yield trials conducted at the International Institute of Tropical Agriculture, Ibadan, Nigeria, indicated that over 3,180 kg/ha/yr of dry cowpea grain can be produced with yields during seasons of adequate rainfall exceeding 2,220 kg/ha (Rachie *et al.*, 1971).

Cowpeas contain an average total protein content of 24 percent, but varieties vary greatly in crude protein and total elemental sulfur (Table 12) (Maner, 1971). Varieties vary from 18 to 29 percent crude protein and, if elemental sulfur can be taken as an estimate of their sulfur amino acid content (Porter, 1972), they can be assumed to vary widely in methionine and cystine.

The essential amino acid content of the cowpea has been studied by several investigators (Jaffé, 1949; Orr and Watt, 1957; Bressani *et al.*, 1961; Elias *et al.*, 1964), and the results indicate that methionine and tryptophan are the most limiting amino acids. Biological

TABLE 12 Distribution and Range of Crude Protein and Sulfur Contents of 141 Varieties of Cowpeas[a]

Protein		Sulfur	
Range (%)	No. of Varieties	Range (% of Protein)	No. of Varieties
18–19	1	0.60–0.70	18
19–20	1	0.70–0.80	54
20–21	2	0.80–0.90	44
21–22	11	0.90–1.00	17
22–23	17	1.00–1.10	6
23–24	25	1.10–1.20	0
24–25	30	1.20–1.30	1
25–26	18	1.30–1.40	0
26–27	21	1.40–1.50	1
27–28	7		
28–29	6		
29–30	2		

[a]From Maner (1971).

TABLE 13 Utilization of Black-eyed Cowpeas by Rats as Affected by Processing and Methionine Supplementation[a]

Treatment[b]	Average Daily Gain[c] (g)	Feed/Gain[c] (kg)
Crude	2.26	3.51
Crude + methionine	3.25	2.94
Cooked	3.84	2.81
Cooked + methionine	4.61	2.48
Soybean meal	4.60	2.72

[a]From Maner and Pond (1971).
[b]Nine rats per treatment.
[c]Values are not significantly different ($P < 0.05$).

studies (Jaffé, 1949; Borchers and Ackerson, 1950; Chavez *et al.*, 1952; Elias *et al.*, 1964) have indicated that cowpea protein is a good source of lysine but a poor source of methionine and cystine. Even though cowpeas are deficient in some of the amino acids, they have a higher nutritive value than common beans (Elias *et al.*, 1964).

The cowpea, like other leguminous seeds, contains a trypsin inhibitor. Studies of the relationship between digestibility of cowpeas and its trypsin inhibitor (Borchers and Ackerson, 1950; Jaffé, 1950) indicate that heating in the autoclave, although sufficient to destroy the inhibitor factor, did not improve the overall digestibility.

Cowpeas contain only small quantities of oxalic acid and, unlike lima beans, only minute quantities of hydrocyanic acid (Oke, 1967).

Early studies with pigs fed cowpeas were conducted by Heitman and Howarth (1960). Cleaned, cull black-eyed peas containing 23.2 percent crude protein were ground without cooking and substituted for barley at levels of 20 and 50 percent in diets for growing pigs. As the percentage of raw black-eyed cowpeas increased, gain in weight decreased. Feed consumption and, apparently, utilization were reduced by the addition of cowpeas, even though no evidence of toxicity was observed.

More recent works with rats (Maner and Pond, 1971) and pigs (Maner, 1971) have indicated that the nutritive value of cowpeas can be greatly increased (Table 13) by cooking and by supplementing with methionine.

Rats fed diets containing cowpeas as the only source of protein grew very poorly when the raw cowpeas were offered (Maner and Pond, 1971). Cooking improved gains by 70 percent and efficiency of feed conversion by about 17 percent. Methionine supplementation improved weight gains of the rats fed both raw or cooked cowpeas. Cooked cowpeas adequately supplemented with methionine have a nutritive value for rats equal to that of soybean meal.

Germinated cowpeas have been utilized to supplement 12 percent protein diets for rats based on *opaque-2* corn (J. H. Maner, unpublished data). Results of this sutdy indicate that although unsupplemented cowpea–*opaque-2* corn diets significantly improved rat gains and feed conversion, performance equal to that obtained with soybean meal–*opaque-2* corn was only obtained after supplementing the diet with 0.05 percent DL-methionine. Other studies (J. H. Maner, unpublished data) indicate that, although methionine supplementation of *opaque-2* corn–cowpea diets is effective in improving performance, no improvement in performance is obtained from methionine supplementation when common corn substitutes for *opaque-2* corn. Although not tested in these studies, these data probably indicate a difference in tryptophan content of the two corns, which becomes limiting in the common corn diet before methionine.

Studies similar to those conducted with rats have also been conducted with growing pigs (Maner, 1971). Pigs fed 16 percent protein diets in which raw or cooked cowpeas provided the only source of protein in the cowpea–sucrose diet, performed very poorly on the raw cowpea diets. Cooking of the cowpeas significantly improved gains, feed consumption, and efficiency of feed conversion. Performance of pigs fed cooked cowpeas was not different from that of pigs fed the 16 percent protein corn–soybean meal control diet. Methionine supplementation was without effect when added to either the raw or cooked cowpea diets (Table 14).

TABLE 14 The Value of Raw and Cooked Cowpeas as a Protein Source for Growing Pigs[a]

Diets	Average Daily Gain (g)	Feed/Gain (kg)
Corn–soybean meal	799	2.53
Raw cowpeas	551	3.43
Raw cowpeas + methionine	483	3.09
Cooked cowpeas	816	2.50
Cooked cowpeas + methionine	815	2.43

[a]From Maner (1971).

Pigeon Pea

The pigeon pea (*Cajanus cajan* or *indicus*) is a shrub, usually cultivated as a perennial and cut back and yields for 2 or 3 years. Easily cultivated and drought resistant, the pigeon pea (also called red gram, dhal, no-eye pea, and guandul) is grown in both the tropics and subtropics. The pigeon pea is used as a human food, livestock feed, and as a green manure crop (Krauss, 1932; Aykroyd and Doughty, 1964; Fike, 1968; Killinger, 1968). The seeds have an acrid taste, which is mainly confined to the seed coat, and are usually eaten when mature and dry. Pigeon peas contain a trypsin inhibitor that can be precipitated with 5 percent trichloroacetic acid and destroyed by moist heat (Pusztai, 1967).

The average total protein of the pigeon pea is 20–23 percent. The amino acid profile of the pea, although adequate in lysine, is very deficient in tryptophan and the sulfur amino acids. According to Orr and Watt (1957), tryptophan is more limiting than methionine. Although pigeon peas contain 0.38 percent oxalic acid, they do not contain any hydrocyanic acid (Oke, 1967).

Although world yields are low, averaging only about 600 kg/ha, yield records from experimental plantings indicate that the potential of the crop is high (Roberts, 1970). Yields in excess of 5,000 kg/ha have been reported in India and in excess of 4,000 kg/ha in Puerto Rico.

Dako (1966) reported that cooking of pigeon peas improved their nutritional value by destroying the substances inhibiting trypsin activity. Raw peas caused hyperactivity of the pancreas, diarrhea, and apathy. Addition of tryptophan or methionine, each alone, to raw or cooked peas had no effect on protein efficiency ratio, but the combination of the two amino acids improved the protein efficiency ratio by about 100 percent.

REFERENCES

Abu-Shakra, S., S. Mirza, and R. Tannous. 1970. Chemical composition and amino acid content of chickpea seeds at different stages of development. J. Sci. Food Agric. 21:91.

Aykroyd, W. R., and J. Doughty. 1964. Legumes in human nutrition. FAO Publ. No. 19. FAO, Rome.

Bajaj, S., O. Mickelsen, L. R. Baker, and P. Markarian. 1971. The quality of protein in various lines of peas. Br. J. Nutr. 25:207.

Balboa, J., E. Zorita, and J. R. Guedas. 1966. Beanmeal (*Vicia faba* L.) as a protein supplement for growing pigs. Rev. Nutr. Anim. Madrid 4:41.

Bandermer, S. L., and R. J. Evans. 1963. Amino acid composition of some seeds. J. Agric. Food Chem. 11:134.

Beeson, W. M., and C. W. Hickman. 1945. Nutritive value of cull peas for fattening hogs. Univ. Idaho Agric. Exp. Sta. Circ. 106.

Bell, J. M. 1965. Growth depressing factors in rapeseed meal. VI. Feeding value for growing–finishing swine of myrosinose-free solvent extracted meal. J. Anim. Sci. 24:1147.

Bell, J. M., and R. J. Belzile. 1965. Goitrogenic properties. *In* Rapeseed meal for livestock and poultry. A review. Publ. 1257. Canadian Department of Agriculture, Ottawa.

Bell, J. M., and A. G. Wilson. 1970. An evaluation of field peas as a protein and energy source for swine rations. Can. J. Anim. Sci. 50:15.

Belzile, R., J. M. Bell, and L. R. Wetter. 1963. Growth depressing factors in rapeseed oil meal. V. The effect of myrosinase activity on the toxicity of the meal. Can. J. Anim. Sci. 43:169.

Borchers, R., and C. W. Ackerson. 1950. The nutritive value of legume seeds. X. Effect of autoclaving and the trypsin inhibitor test for 17 species. J. Nutr. 41:339.

Bowland, J. P. 1965. Feeding value of rapeseed meal for swine. *In* Rapeseed meal for livestock and poultry. A review. Publ. 1257. Canadian Department of Agriculture, Ottawa.

Bowland, J. P., and J. F. Standish. 1966. Growth, reproduction, digestibility, protein and Vitamin A retention of rats fed solvent-extracted rapeseed meal or supplemental thiouracil. Can. J. Anim. Sci. 46:1.

Bowman, D. E. 1944. Fractions derived from soybeans and navy beans which retard tryptic digestion of casein. Proc. Soc. Exp. Biol. Med. 57:139.

Brambila, S., and A. S. Shimada. 1967. Unpublished data. Cited in A. S. Shimada and S. Brambila. 1967. Actividad antiproteolitica del garbanzo. Tec. Pecu. Mex. 10:5.

Bravo, F. O., and E. Cabello. 1969. The effect of three combinations of safflower cake and molasses in finishing rations for pigs. Tec. Pecu. Mex. 11:38.

Bressani, R., L. G. Elias, and D. A. Navarrette. 1961. Nutritive value of Central American beans. IV. The essential amino acid content of samples of black beans, red beans, rice beans, and cowpeas of Guatemala. J. Food Sci. 26:525.

Bressani, R. 1969. Variación en el contenido de nitrógeno, metionina, cistina y lisina de selecciones de fríjol. *In* C. L. Anas [ed.] Programa cooperativo Centroamericano para el mejoramiento de cultivos alimenticios. Fríjol XI Reunión Anual. Publ. Misc. No. 68. San Salvador, El Salvador.

Butterworth, M. H., and H. C. Fox. 1963. The effect of heat treatment on the nutritive value of coconut oil meal and the prediction of nutritive value by chemical methods. Br. J. Nutr. 17:445.

Catron, D. V., and V. W. Hays. 1958. La torta de soya en la moderno nutricion animal. Boletin de soybean Council of America, Rome, 35 p.

Chavez, N., N. R. Teodosio, A. Gomex de Matos, Jr., C. A. Lima, and J. L. de Almuda. 1952. The protein of the cowpea (*Vigna sinensis*) in nutrition. Rev. Brasil. Med. 9:603.

Clandinin, D. R. 1965. Feeding value of rapeseed meal for poultry. *In* Rapeseed meal for livestock and poultry. A review. Publ. 1257. Canadian Department of Agriculture, Ottawa.

Clarke, H. E. 1970. The evaluation of the field bean (*Vicia faba* L.) in animal nutrition. Proc. Nutr. Soc. 29:64.

Conner, J. K., A. R. Neill, and H. W. Burton. 1971. Navy beans (*Phaseolus vulgaris*) as a protein source in larger diets. Aust. J. Exp. Agric. Anim. Husb. 11:387.

Cornelius, J. A., and W. D. Raymond. 1967. Some oilseeds from tropical herbaceous crops. Trop. Sci. 9:75–89.

Creswell, D. C., and C. C. Brooks. 1971a. Composition, apparent digestibility and energy evaluation of coconut oil and coconut meal. J. Anim. Sci. 33:366.

Creswell, D. C., and C. C. Brooks. 1971b. Effect of coconut meal on *Coturnix* quail and of coconut meal and coconut oil on performance, carcass measurements and fat composition in swine. J. Anim. Sci. 33:370.

Cunha, T. J., E. J. Warwick, M. E. Ensminger, and N. K. Hart. 1948. Cull peas as a protein supplement for swine feeding. J. Anim. Sci. 7:117.

Dako, D. H. 1966. The protein value of African legumes in relation to pretreatment and combination with other foods. PhD Thesis. Liebig-Univ., Giessen [Nutr. Abstr. Rev. 1968. 38(2630):469].

Delic, I., T. Bokorov, A. Sreckouic, and M. B. Nikolic. 1963. Biological value of sunflower oilmeal as a protein feed for fattening pigs. Stocarstvo 17:464. (Nutr. Abstr. Rev. 1964. 34:596).

Delic, I., M. Nicolic, S. Sargin, and T. Bokorov. 1964. A combined protein supplement of sunflower oil meal and lucerne leaf meal to replace soya bean oil meal in mixtures for fattening pigs. Veterinaiya Sarajevo 13:195. (Nutr. Abstr. Rev. 1965. 35:849).

Devilat, J. 1965. El afrecho de raps como suplemento proteínico en raciones de cría y engorda de cerdos. Invest. Ganad. Chile 1:61.

Downey, R. K. 1965. Rapeseed botany, production and utilization. *In* Rapeseed meal for livestock and poultry. A review. Publ. 1257. Canadian Department of Agriculture, Ottawa.

Eden, A. 1968. A survey of the analytical composition of field beans (*Vicia faba* L.). J. Agric. Sci. Camb. 70:299.

Elias, L. G., R. Colindres, and R. Bressani. 1964. The nutritive value of eight varieties of cowpeas (*Vigna sinensis*). J. Food Sci. 29:118.

Esnaola, M. A., and J. A. Ochoa. 1970. Uso de afrecho de raps en raciones de crianza y engorda de cerdos. Agric. Tec. 30:90.

FAO. 1970. Production yearbook. Vol. 24. FAO, Rome.

Fike, W. T. 1968. Pigeon peas (*Cajanus cajan*), a new green manure crop for the South. Mimeograph. North Carolina State University, Raleigh.

Gallo, J. T., and J. H. Maner. 1970. La torta de ajunjolí para cerdos en crecimiento y acabado. Revista ICA 5:107.

Goatcher, W. D., and J. McGinnis. 1972a. Effect of autoclaving beans, and amino acid and antibiotic supplementation upon performance of chicks fed one of four varieties of dry beans. Poult. Sci.

Goatcher, W. D., and J. McGinnis. 1972b. Influence of beans, peas and lentils as dietary ingredients on the growth response of chicks to antibiotic and methionine supplementation of the diet. Poult. Sci. 51:440.

Grieve, C. M., D. F. Osbourn, and F. O. Gonzáles. 1966. Coconut oil meal in growing and finishing rations for swine. Trop. Agric. Trin. 43:257.

Heitman, H., Jr., and J. A. Howarth. 1960. Black-eyed peas as a swine feed. J. Anim. Sci. 19:164.

Henry, Y. 1970. La feverole dans l'alimentation du porc. Bull. Tech. Inf. 253:1.

Hervas, E., J. Viteri, and J. H. Maner. 1965. Unpublished data, Instituto Nacional de Investigaciones Agropecuaria. Quito, Ecuador.

Hussar, N., and J. P. Bowland. 1959. Rapeseed oil meal as a protein supplement for swine and rats. I. Rate of gain, efficiency of food utilization, carcass characteristics and thyroid activity. Can. J. Anim. Sci. 39:84.

Jacob, E. 1967. Contenido de ácido cianhídrico en porotas (*Phaseolus vulgaris*) crudo y precocido, de las variedades de mayor consumo en Chile. Nutr. Bromatol. Toxicol. 6:75.

Jaffé, W. G. 1949. Limiting essential amino acids in some legume seeds. Proc. Soc. Exp. Biol. Med. 71:469.

Jaffé, W. G. 1950. Protein digestibility and trypsin inhibitor activity of legume seeds. Proc. Soc. Exp. Biol. Med. 75:219.

Johns, C. D., and A. J. Finks. 1920. Studies in nutrition. 2. The role of cystine in nutrition as exemplified by nutrition experiments with the proteins of navy beans (*Phaseolus vulgaris*). J. Biol. Chem. 41:379.

Johnson, R. H., and W. D. Raymond. 1964. The chemical composition of some tropical food plants. III. Sesame seed. Trop. Sci. 6:173.

Kakade, M. L., and R. J. Evans. 1964. Effect of methionine vitamin B12 and antibiotic supplementation on protein nutritive value of navy beans. Proc. Soc. Exp. Biol. Med. 115:890.

Kakade, M. L., and R. J. Evans. 1965. Nutritive value of navy beans (*Phaseolus vulgaris*). Br. J. Nutr. 19:269.

Kakade, M. L., and R. J. Evans. 1966. Growth inhibition of rats fed raw navy beans (*Phaseolus vulgaris*). J. Nutr. 90:191.

Kakade, M. L., and R. J. Evans. 1969. Unavailability of cystine from trypsin inhibitor as a factor contributing to the poor nutritive value of navy beans. J. Nutr. 99:34.

Kakade, M. L., J. E. Smith, and R. Barchers. 1968. Effect of navy bean fraction on protein synthesis in rats. Proc. Soc. Exp. Biol. Med. 128:811.

Kelly, J. F. 1971. Genetic variation in the methionine levels of mature seeds of common beans (*Phaseolus vulgaris* L.). J. Am. Soc. Hortic. Sci. 96:561.

Killinger, G. B. 1968. Pigeon peas [*Cajunus cajan* (L.) Druce], a useful crop for Florida. Proc. Soil Crop Sci. Soc. Fla. 28:161.

Kratze, F. H., and D. E. Williams. 1951. Safflower oil meal in rations for chicks. Poult. Sci. 30:417.

Krauss, F. G. 1932. The pigeon pea (*Cajanus cajan*), its improvement, culture and utilization in Hawaii. Hawaii Agric. Exp. Sta. Bull. 64.

Kroening, G. H. 1968. Protein quality studies with swine. Unpublished paper. Washington State University, Pullman.

Lehrer, W. P., Jr., and C. W. Hodgson. 1946. Cull peas for fattening hogs in drylot. Idaho Agric. Exp. Sta. Circ. 107.

Leleji, O. I. 1971. The genetics of crude protein and its relation to physiological and agronomic factors in dry beans. PhD thesis. Cornell University, Ithaca, New York.

Livingstone, R. M., V. R. Fowler, and A. A. Woodham. 1970. The nutritive value of field beans (*Vicia faba* L.) for pigs. Proc. Nutr. Soc. 29:46A.

Lobban, J. M. 1963. Peas as feed for pigs. Trials show profitable level of field pea rations. J. Agric. Sci. Aust. 66:462.

Loosli, J. K., J. O. Pena, L. A. Ynalues, and V. Villegas. 1954. The digestibility by swine of rice bran, copra meal, coconut meat, coconut residue and two concentrate mixtures. Philipp. Agric. 38:191.

Luscombe, 1969. Cited by Clarke, 1970.

Maner, J. H. 1971. Swine production systems. Centro Internacional de Agricultura Tropical Annual Report. Cali, Colombia.

Maner, J. H., and J. T. Gallo. 1963. Valor nutritivo de la torta de ajonjolí como reemplazo de la torta de soya en dietas para cerdos en crecimiento y acabado. Primer Congreso Nacional de la Industria Porcina, Bogatá, Colombia. p. 1–3.

Maner, J. H., and J. T. Gallo. 1970. La torta de ajonjolí en alimentación de cerdos. II. Efecto de la suplementación de las tortas de soya y ajonjolí (sesame) con metionina, en raciones para cerdos. Revista ICA 5:113.

Maner, J. H., and W. G. Pond. 1971. Effect of processing and methionine supplementation on the utilization of black-eyed peas (*Vigna sinensis*) by rats. J. Anim. Sci. 33:233.

Manns, J. G., and J. P. Bowland. 1963. Solvent-extracted rapeseed oil meal as a protein source for pigs and rats. I. Growth, carcass characteristics and reproduction. Can. J. Anim. Sci. 43:252.

Mitra, R. C., and P. S. Misro. 1967. Amino acid of processed proteins. J. Agric. Food Chem. 15:697.

Moran, E. T., Jr., J. D. Summers, and G. E. Jones. 1968. Fieldpeas as a major dietary protein source for the growing chick and laying hen with emphasis on high-temperature steam pelleting as a practical means of improving nutritional value. Can. J. Anim. Sci. 48:47.

Navratil, B. 1965. Feeding value of horse beans for fattening pigs. Zivoc. Byr. 10:899. [Nutr. Abstr. Rev. 36(7117):1163].

Oke, O. L. 1967. Chemical studies of some Nigerian pulses. West Afr. J. Biol. Appl. Chem. 9:52.

Orr, M. L., and B. K. Watt. 1957. Amino acid contents of foods. Home Economics Research Department, U.S. Department of Agriculture, Rep. No. 4. Washington, D.C.

Pearson, A. M., H. D. Wallace, and J. F. Hentges, Jr. 1954. Sunflower seed meal as a protein supplement for beef cattle and swine. Univ. Fla. Agric. Exp. Sta. Bull. 553.

Porter, W. 1972. Genetic control of protein and sulfur contents in dry beans, *Phaseolus vulgaris.* PhD thesis. University of Purdue, Lafayette, Indiana.

Pennacchiotti, M. I., and H. Schmidt-Hebbel. 1966. Amino acids in Chilean legumes. Arch. latinoam. Nutr. 18:233.

Purdom, M. E., and R. V. Brown. 1967. Biological response of rats fed amino acid supplemented pea bean (*Phaseolus vulgaris*) diets. Arch. latinoam. Nutr. 17:117.

Pusztai, A. 1967. Trypsin inhibitors of plant origin, their chemistry and potential role in animal nutrition. Nutr. Abstr. Rev. 37:1.

Rachie, K. O., P. N. Mehta, and M. A. Akinpela. 1971. Grain legume improvement program annual report–1971. International Institute of Tropical Agriculture, Ibadan, Nigeria.

Renner, R., D. R. Clandinin, A. B. Morrison, and A. R. Rabblee. 1953. The effect of processing temperatures on the amino acid content of sunflower seed oil meal. Poult. Sci. 32:922.

Ries, S. K. 1971. The relationship of protein content and size of bean seed with growth and yield. J. Am. Soc. Hortic. Sci. 96:557.

Rigas, D. A., and E. E. Osgood. 1955. Purification and properties of the phyto hemoglutinin of *Phaseolus vulgaris*. J. Biol. Chem. 212:607.

Robblee, A. R., 1965. Status of rapeseed meal as a protein supplement. *In* Rapeseed meal for livestock and poultry. A review. Publ. 1257. Canadian Department of Agriculture, Ottawa.

Roberts, L. M. 1970. The food legumes. Recommendations for expansion and acceleration of research to increase production of certain of these high-protein crops. Mimeograph. The Rockefeller Foundation, New York.

Rosa, J. G., J. J. Romero, A. Skoknic, and J. Devilat. 1969. Efecto de distintos suplementos protéicos y de la forma de presentación del grano de maíz en la alimentación de cerdos en crianza-engorda, proporcionados a libre elección. Agric. Tec. Chile 29:133.

Rutger, J. N. 1970. Variation in protein content and its relation to other characters in beans (*Phaseolus vulgaris* L.). Paper given at the Tenth Dry Bean Research Conference, Davis, California, Aug. 12–14, 1970.

Shimada, A. S., and A. Aguilera. 1966. Utilization of safflower meal in the feeding of growing pigs. Tec. Pecu. Mex. 7:6.

Shimada, A. S., and S. Brambila. 1966. Evaluation of the substitution of soybean meal with cottonseed meal and safflower meal in corn based diets with and without molasses for growing–finishing pigs. Tec. Pecu. Mex. 8:30.

Shimada, A. S., and S. Brambila. 1967a. El valor nutritivo del garbanzo forrajero (*Cicer arietinum* L.) como fuente de energía y proteína para el cerdo. Tec. Pecu. Mex. 9:27.

Shimada, A. S., and S. Brambila. 1967b. Efecto del cocimiento del garbanzo (*Cicer arietinum* L.) sobre su valor nutritivo para cerdos. Tec. Pecu. Mex. 10:5.

Shimada, A. S., and S. Brambila. 1967c. El efecto de substituir harina de pescado por garbanzo en raciones a base de sorgo y pasto de algodón para el cerdo en crecimiento. Tec. Pecu. Mex. 10:19.

Silbernagel, M. L. 1970. Bean protein improvement work by USDA–Bean and pea investigation. Paper presented at the Tenth Dry Bean Research Conference, Davis, California. Aug. 12–14, 1970.

Smith, K. J. 1968. A review of the nutritional value of sunflower meal. Feedstuffs 40:20.

Tannous, R. I., and M. Ullah. 1969. Effects of autoclaving on nutritional factors in legume seeds. Tropic. Agric. Trin. 46:123.

Thomke, S., and A. Frölick. 1968. Utfodringsförsök med rostode ärter till slaktsuin. Series A-NR98. Report of Agric. Col. of Sweden, Uppsala, p. 20–26.

Tkacev, I. F., G. A. Taranenko, A. P. Bacikalo, V. I. Zvjagincev, Ju. P. Macuk, and Ju. T. Kouazmyha. 1965. Feeding value of sunflower oilmeal prepared by different processes. Vestu. sel'skohoz Nauk. No. 5. p. 119–128. (Nutr. Abstr. Rev. 1966. 36:244)

Tkacev, I. F., G. A. Taranenko, V. I. Zvjagivcev, and A. P. Vacikalo. 1964. Feeding value of sunflower oilmeal with different degrees of temperature denaturation of the protein. Zivotnovodstvo No. 7. p. 74–77. (Nutr. Abstr. Rev. 1965. 35:207)

Warwick, E. J., T. J. Cunha, and M. E. Ensminger. 1948. Cull peas for growing and fattening swine. Wash. Agric. Exp. Sta. Bull. No. 500.

Wetter, L. R. 1965. Chemical composition of rapeseed meal. *In* Rapeseed meal for livestock and poultry. A review. Publ. 1257. Canadian Department of Agriculture, Ottawa.

Young, C. G. 1965. Processing of rapeseed meal. *In* Rapeseed meal for livestock and poultry. A review. Publ. 1257. Canadian Department of Agriculture, Ottawa.

Zaghi, S., and R. Bressani. 1969. Uso de recursos alimenticios Centroamericanos para el fomento de la industria animal. II. Composición química de la semilla y de la harina de torta de ajonjolí. (*Sesamun indicum*). Turrialba 19:34.

M. A. Steinberg

TECHNOLOGICAL DEVELOPMENTS IN FISH PROCESSING AND IMPLICATIONS FOR ANIMAL FEEDING

This paper deals with certain data on the amount and sources of whole fish protein available in the United States for animal feeding, some suggestions as to how the supply may be affected in the future by technological developments for improving the utilization of our fishery resources, and, finally, some views on existing sources of unutilized fish protein and the likelihood of their contributing to meeting the protein needs of the animal industry.

FISH PROTEIN SUPPLY

It is necessary first to put the industrial fishery—that is, that fishery whose product is not intended for direct human consumption—into perspective with the entire U.S. fishery. In 1971, the per capita consumption of edible fishery products was approximately 11 lb. This figure has remained at about that level for the past 20 yr (Riley, 1971). The amount of fish available as fish meal was equivalent to 57 lb of liveweight fish/person. The total industrial fishery volume thus exceeds that of the food-fish industry by more than five times.

The bulk of domestically produced fish meal comes from menhaden.

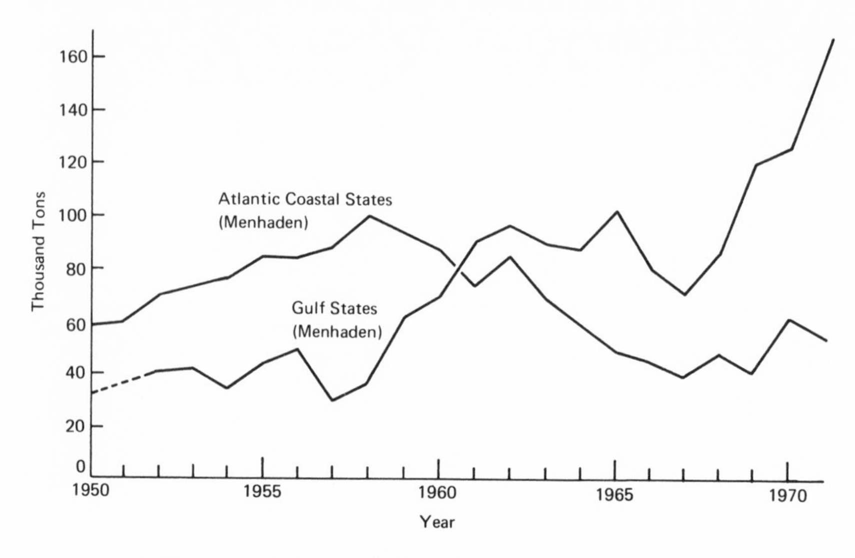

FIGURE 1 Annual U.S. production of menhaden fish meal, 1950–1971.

Other contributing species are tuna, mackerel, herring, and anchovy. Anchoveta, taken from the coastal waters of Peru, are at least equal in importance to menhaden as a raw material for the production of fish meal marketed in the United States.

Figure 1 shows the change in the source of domestically caught menhaden from 1950 to the present. It is clear that the Atlantic Coast states, once the major domestic contributors to our industrial fishery, have declined as a source of menhaden since the late 1950's, largely as a result of overfishing. The Gulf states have shown a more than corresponding increase in the production of menhaden.

Figure 2 shows the annual production of fish meal in the United States, Peru, and on a worldwide basis from 1955 to 1970. On this scale, the production in the United States is seen to be relatively small and stable. The steep rise in world production since 1958 is only in part explained by increased Peruvian production. World production is increasing at a faster rate than is Peruvian production.

In Figure 3, we see that from 1950 to approximately 1960, 75–80 percent of the total available fish meal in the United States was from domestic production. Beginning about 1963, imported meal comprised more than 50 percent of the total amount of available fish meal in the United States. This situation continued until 1969–1970. The reason for this sudden and large demand for fish meal is somewhat difficult to explain without a study of the costs and availability of protein that is

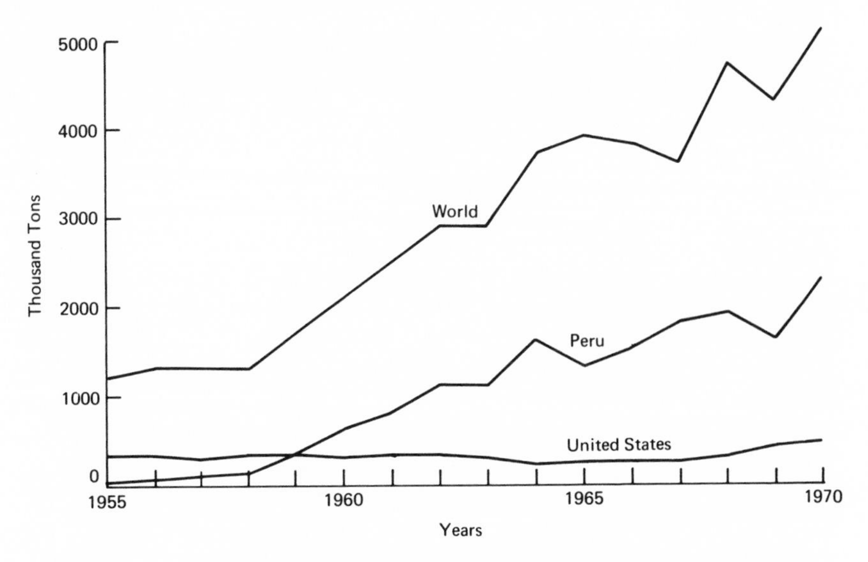

FIGURE 2 Annual U.S., Peruvian, and world production of fish meal, 1955–1970.

competitive with fish meal, but it is of interest that, in mid-year, the Peruvian government sharply curtailed the harvesting of anchoveta in an effort to protect the resource from overfishing. This is not the first time that this has occurred, and it indicates that the Peruvian anchoveta is as vulnerable to overfishing as are other species.

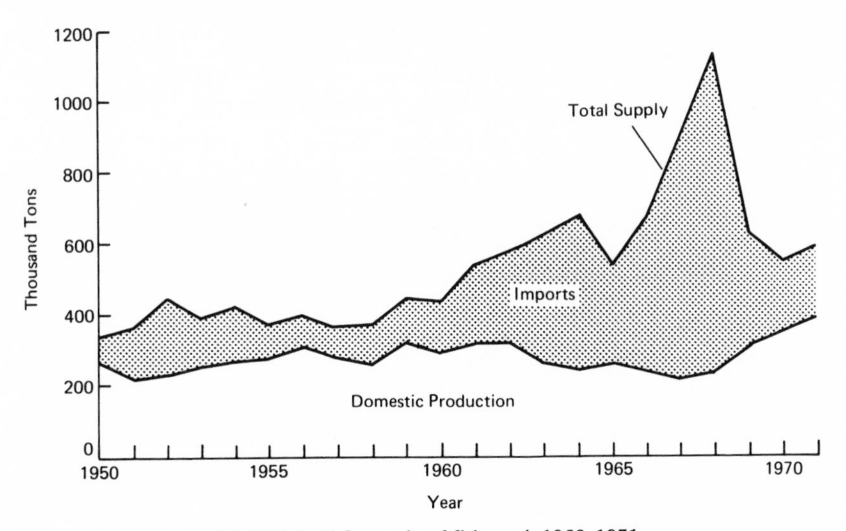

FIGURE 3 U.S. supply of fish meal, 1950–1971.

Since 1950, the 10-year average supply of fish meal in the United States has risen from about 250,000 tons (1950–1959) to about 400,000 tons (1960–1969).

TECHNOLOGICAL DEVELOPMENTS

The United States imports approximately 300 million lb of fish-fillet blocks per year [National Marine Fisheries Service (NMFS), 1968]. These come from such countries as Iceland, Canada, West Germany, South Africa, and Argentina. The cost of these products is increasing rapidly and may, very soon, if the resource from which they are manufactured continues to decline, cease to be competitive with other animal-protein food products. The following is quoted from the U.S. Department of Agriculture report *National Food Situation*, May 1972:

> Declining availability of cod, the traditional favorite in sticks and portions, led to sharply rising raw material costs. Consumer resistance to price increases led sticks and portions producers to cut back production for the first time since 1967.

The technology may soon be available to permit using such species as menhaden for processing into a modified fish block that can then be processed further into fish sticks and fish portions. If we assume that 50 percent of the presently imported blocks could be replaced by comminuted fish blocks made from species such as menhaden, the total tonnage of fish required to meet the demand would be about 156,000 tons/yr. This is equivalent to approximately 30,000 tons of meal, about 8 percent of domestic production in 1971.

Early in the 1960's, the then Bureau of Commercial Fisheries began to develop a process for the production of fish-protein concentrate for use in human feeding. Up to the present, developmental work has been done only with lean fish—those species having a fat content of no more than 3 or 4 percent. Work is just now under way to modify the process to use fatty species (menhaden, anchovy, herring, etc.) as raw material. In fact, the first two of these may be the only single-species resources available in quantity off the coast of the United States for processing into fish-protein concentrate. It is impossible to estimate what the domestic market will be for fish-protein concentrate—as it is now being developed, it is a nutritional additive and has no other function. One can get a rough idea, however, of what the market must be if one refers to production capabilities. Minimum capacity for a profitable plant has been estimated to be 200 tons of fish/day. On an annual basis, one plant operating on an 8-hr day for 240 days would produce fish-protein concentrate equivalent in tonnage to about 10,000 tons of fish meal. Of course, the multiplier for this quantity depends on how the food-processing industry views it as a protein supplement. And industry's view, in turn, depends upon whether the consumer wants protein-supplemented foods sufficiently to pay the costs of supplementation. In any case, part or all of the menhaden resource could be diverted to the manufacture of fish-protein concentrate. The critical factor, demand, has yet to be determined.

Another example of the potential use of fish-muscle protein was provided by the poultry industry a few years ago. The selling of prepackaged poultry parts results in an overabundance of backs, necks, and wings. These parts are high in fat, skin, and bone and relatively low in muscle tissue. In looking for a market for these parts, the poultry industry obtained approval from the Department of Agriculture to use poultry as an ingredient in sausage products to the extent of 15 percent of all ingredients, minus water. Higher levels can be used, but special labeling requirements have to be met. Because of the relatively poor functional properties of the mixture of skin, fat, and muscle, the meat industry has not embraced the idea of using poultry to any significant extent in their sausage products. However, some people in the NMFS recognize that the meat industry might provide considerable potential for the use of fishery products presently of low value—for example, menhaden. Several tests have been carried out with the muscle of menhaden and with species of fish commonly taken from the waters of the northeastern Pacific to determine how well they perform as partial replacers of the lean beef used as emulsifiers and nutritional protein sources in sausage products. This work has been done in cooperation with commercial sausage manufacturers, and fish muscle has been found by these food processors to be eminently satisfactory in terms of function. Products made include frankfurters, bologna, pork sausage, and luncheon meats. Texture, flavor, yield, and resistance to spoilage have all been found to be satisfactory when 15 percent of the lean-meat portion of the product was replaced by an equal weight of fish flesh. The sausage manufacturing industry in 1969 was large enough to absorb 200 million lb of fish muscle if used at the 15 percent level in 50 percent of the product. This is equivalent to 400 million lb of whole fish or about 40,000 tons of meal/yr, approximately 10 percent of 1971 domestic production. The present definition of sausages and sausage products precludes the use of fish. We are hoping that it will be changed, because of the benefits that fish in these products can bring to the consumer, the meat industry, and the fish industry.

Studies on the properties of fish-protein concentrate have taken us to

TABLE 1 Functional Properties of Some Proteins and Types of Products in Which They are Used[a]

Property	Product Type
Emulsification	Salad dressings
	Sausage products
	Baked goods
	Frozen desserts
Fat absorption	Sausage products
Fat absorption control	Fried pastries
Water absorption	Baked goods
	Pasta products
Cohesion	Processed meats and sausage products
	Minced fishery products
Texture	Simulated meats
	Fish pastes
Whippability	Whipped desserts
	Toppings
	Baked goods

[a]From Hammonds and Call (1970).

the investigation of fish muscle protein as a source of functional protein isolates. Examples of functionality are shown in Table 1. The ability to form stable emulsions, absorb fat, limit the absorption of fat, absorb water, cause discrete muscle particles to hold together, give plasticity, and incorporate and hold air, as in a foam, are typical, frequently characteristic functional properties of proteins, for which these materials have considerable value to the food manufacturing industry. Without the emulsifying property of egg yolk, salad dressings and mayonnaise would not have their characteristic stability. The myofibrillar protein of beef and pork is responsible for forming the stable emulsions essential to the physical properties of sausage products. Dry milk solids and sodium caseinate, as well as casein, form and stabilize the emulsions in baked products and in frozen, whipped dessert toppings. Usually, functional proteins are effective in relatively small amounts, but contribute little to the nutritional property of the product to which they are added. There are exceptions, however. The myofibrillar protein of meat is the primary protein in sausage products. Without it, frankfurters and other sausage products would contain little more than fat and water.

Table 2 shows costs of presently used functional proteins, their composition in terms of percent protein, their bulk price, and the cost per pound of protein. I have added fish meal to the list, assuming a protein

TABLE 2 Cost of Selected Protein Ingredients, 1970[a]

Ingredient	Percent Protein	Bulk Price (¢/lb)	Cost/lb of Protein (¢/lb)
Nonfat dry milk	36	20.0– 25.0	55.6– 69.4
Dry whole milk	26	40.0– 45.0	153.8–173.1
Soy proteins			
flour	53	6.7– 6.8	12.6– 12.8
grits	53	6.5– 6.8	12.3– 12.8
concentrates	70	19.0– 24.0	27.1– 34.3
isolates	95	35.0– 40.0	36.8– 42.1
Sodium caseinate	95	80.0–100.0	84.2–105.3
Cottonseed flour	55	10.0– 12.0	18.2– 21.8
Fish meal	65	7.5– 9.0	11.5– 14.0

[a]From Hammonds and Call (1970).

content of 65 percent and a price range of $150–180/ton. This gives us a cost per pound of protein in fish meal of 11.5–14 cents. This is comparable in protein cost only to such low-cost functional proteins as soy protein flour and soy grits. I am not attempting here to compare nonfunctional fish meal, whose primary use is as animal feed, with functional proteins intended for use in foods. I am simply trying to establish that the fish meal manufacturer receives a price for his product that is not at all attractive when compared with the price received by the processors of other proteins, the value of which is determined not by their amino acid composition or their nutritional value but rather by the particular functions that they perform when added as ingredients to processed foods. This, then, is the crux of the argument: If the fish meal manufacturer can sell for a higher price the protein component of the raw material he now converts into fish meal, he will do so. This should be an attractive opportunity for a segment of the fish industry whose contribution to total U.S. landings is nearly 40 percent, but whose ex-vessel return is less than 6 percent of the value of U.S. landings.

Table 3 lists some common food types that include functional proteins as an essential constituent and shows the amount of functional proteins used in the manufacture of these products in 1969 and projects this use to 1980, at which time it is assumed that approximately 1.7 billion lb of protein will be used in the products listed here. This table simply illustrates the ubiquity of functional proteins and the certainty of their increased use. It obviously does not, and cannot, take into account food products that do not now exist and the dependence of these products on proteins that perform functions necessary to provide the

TABLE 3 Selected Protein Growth for Selected Product Categories[a]

Product	Annual Growth Rate (%)	Protein (Millions of lb)	
		1969	1980
Baby food	1.0	3.5	3.9
Baked goods and baking needs			
snack food	6.0	10.0	19.0
all other	1.5	91.0	107.1
Breakfast food			
instant breakfast	8.0	12.8	29.8
all other	1.7	5.1	6.2
Candy	3.0	16.6	23.0
Canned and processed meat	9.3	92.2	245.9
Coffee whitener	6.0	12.0	22.8
Dairy products			
imitation milk	–	–	188.0
synthetic ice cream	5.0	3.8	6.5
all other	1.0	98.1	109.1
Desserts and toppings	6.0	31.7	60.0
Diet drink	2.0	8.4	10.5
Frozen food	3.6	3.8	5.6
Macaroni/pasta products	3.0	1.5	2.1
Pet food	5.4	230.9	426.0
Soup	0.0	1.5	1.5
SUBTOTAL		621.3	1,267.0
All other uses	6.7	207.3	420.0
TOTAL		828.6	1,687.0

[a]From Hammonds and Call (1970).

characteristic properties of these products, whatever they may be.

This table shows such food types as coffee whiteners, dairy products, dessert toppings, diet drinks, and juices, all of which seem rather unlikely as uses for fishery products. But, in fact, we are now in the advanced stages of development of two modified fish-muscle proteins—one, a chemically derived succinylated myofibrillar protein and the second, an enzymically altered myofibrillar protein—that are essentially flavorless and odorless, are soluble in water, and are extremely effective as emulsifying agents. One of them is also a whipping agent and is stable to heat. Either of them would be useful in most of the products that I have just named. One of these proteins has been shown to be effective as a 30 percent replacement for egg white in the preparation of angel food cake and sponge cake. This same protein has been used as a complete replacement for whole egg in the preparation of oatmeal pastry bars.

If we assume that these protein isolates and others derived from fish muscle will satisfy 10 percent of the projected 1980 market, we find that we are talking about the equivalent of approximately 180,000 tons of fish meal/yr. This is about 50 percent of the meal produced in the United States in 1971.

Economics and technology will ultimately determine how the raw material supply is used. One thing seems certain. Other markets will compete intensely with the animal feed market for fish protein. Other markets have the competitive advantage of selling their products at a higher price and higher profit margin than that for fish meal.

Technology is already making itself felt in its effect on animal feeding of fish protein. A solvent-extracted, low-fat, high-protein product made from herring is now available commercially for animal feeding. The product, intermediate between fish meal and fish-protein concentrate, seems to be waiting for clarification of marketing restrictions on fish-protein concentrate before taking the final step into the food market.

Production of each of the type of product that have been discussed, with the exception of fish-protein concentrate, requires the removal of sarcoplasmic protein. The proteins in this group comprise about 25 percent of the total protein content of fish muscle, but most removal procedures account for only a little more than one half of that. This protein can be recovered quantitatively from wash water as a phosphate complex and has a PER of 4.0. The PER of the Animal Nutrition Research Council (ANRC) reference casein, determined at the same time, is 3.11 (Spinelli and Koury, 1970). Based on the estimated potential use for minced fish blocks, sausage products, and protein isolates, approximately 16,000 tons of protein could become available annually for use in animal feeds. This is equivalent to 17 percent of our present annual production of fish meal containing 60 percent protein.

ALTERNATIVE SOURCES

It is useful to examine the criteria that must be met by a resource before it is suitable for use in the industrial fisheries. First, it must be abundant, and, related to this, it must also be concentrated. The combination of abundance and concentration directly affects the catch rate. The price per pound of these fish, which is, of course, a factor in determining the selling price, is—at least in part—determined by the cost of catching, and that is—in large part—a function of catch rate. The resource must also be reasonably close to shore. This seems obvious as a requirement, but an example of failure to consider this factor has come up in discussions in which antarctic krill have been suggested as an alternative source of protein for animal feeding. Another factor that must be considered is the desirability for use as food of the species under consideration. The point of this discussion has been to emphasize that use as a food product is more profitable than use as an animal-feed product and that a resource suitable for use as human food is priced out of the animal-feed market. Finally, the resource must not suffer from social constraints. To be specific, the catch of the abundant, concentrated, close-in anchovy resource off the coast of California is limited not by the ability of the fisherman to take these fish but rather by quotas set by the state and determined at least partly by the sport fishermen, who are of the opinion—right or wrong—that these fish are important to the welfare of sport fish that feed on the anchovies. Thread herring, which are relatively abundant off the west coast of Florida, cannot be fished because sport fishermen there consider that these fish are essential to the sport fisheries. This is true of the herring resource found off the coast of Alaska. In this case, the commercial fisherman want the resource to be protected in order to provide food for the large stocks of salmon that return annually to Alaska waters. If these constraints are lifted, these resources may become available for manufacture into fish meal. Even then, however, we have the possibility of fish-protein concentrate, rather than fish meal, as an end product. This, of course, as indicated above, depends upon the market that develops for fish-protein concentrate.

Recovery of wastes from fish-processing plants in an effort to comply with federal regulations concerning pollution of the environment could represent a source of low-protein (approximately 40 percent), high-ash fish meal for use in animal feeds. The city of Kodiak, Alaska, probably contains the most concentrated collection of fish-processing plants in the United States. The plants in this city now handle about 100 million lb of fish and shellfish/yr (data from Environmental Protection Agency, Seattle, 1972). The products range from crab, shrimp, and salmon to small scrap fish caught incidental to other operations. All wastes, including crab and shrimp shells and whole scrap fish, would, if converted to meal, yield approximately 20 million lb of meal/yr. Because of the concentration of processing plants, this area is probably the only one in Alaska in which a reduction operation might be economically feasible, and even here the operation would probably not be profitable unless special uses could be found for the chitinous material contained in the crab and shrimp shell. The tuna industry is also sufficiently concentrated in the United States to permit reduction of waste to meal, but this industry has been doing this for some years and therefore does not represent a new source of fish meal. Most other fish-processing operations on the Pacific Coast are so widely scattered and relatively small at any particular location as to preclude the profitable conversion of wet wastes to meal. This is generally characteristic of the fishing industry. Where there are exceptions, such as at the Boston Fish Pier, fillet wastes are already converted to fish meal or, as is characteristic of the Gulf of Mexico processors, the largest of the processing plants are those that process shrimp, whose waste material is highest in shell and lowest in economically recoverable protein.

There are serious doubts concerning the future use of latent species of fish for industrial purposes. The recovery of processing wastes for conversion to meal, while promising in the case of sarcoplasmic proteins, is inadequate to meet the needs of the animal industry.

REFERENCES

Food and Agricultural Organization. 1961. Yearbook of fishery statistics, 1960. Vol. 12. FAO, Rome. 419 p.

Food and Agricultural Organization. 1966. Yearbook of fishery statistics, 1965. Vol. 21. FAO, Rome. 366 p.

Food and Agricultural Organization. 1971. Yearbook of fishery statistics, 1970. Vol. 31. FAO, Rome. 320 p.

Hammonds, T. M., and D. L. Call. 1970. Utilization of protein ingredients in the U.S. food industry. Part II. The future market for protein ingredients. Agricultural Economics Research Rep. 321. N.Y. State College of Agriculture, Ithaca. 36 p.

Kolhonen, J. A. 1972. Current economic analysis 1-17. Industrial fishery products situation and outlook—1971 annual review. National Marine Fisheries Service, U.S. Department of Commerce, Washington, D.C. 35 p.

National Marine Fisheries Service. 1971. Fishery statistics of the United States, 1968. U.S. Department of Commerce, Washington, D.C. 578 p.

Riley, F. 1971. Current fishery statistics No. 5600. Fisheries of the United States, 1970. U.S. Department of Commerce, Washington, D.C. 79 p.

Spinelli, J., and B. Koury. 1970. Phosphate complexes of soluble proteins. Agric. Food Chem. 18:284–288.

U.S. Department of Commerce. 1971. Current fisheries statistics No. 5515. Fish meal and oil—1970. Annual summary. Washington, D.C. 4 p.

J. M. Asplund and W. H. Pfander

PRODUCTION OF SINGLE-CELL PROTEIN FROM SOLID WASTES

The industrialized and affluent societies of North America and western Europe annually discharge millions of tons of dry matter as byproducts of industry. These wastes represent a serious loss of oxidizable carbon and hydrogen at a time when energy and food supply are approaching a crisis. As point sources, they also constitute an almost overwhelming source of environmental pollution. The obvious course of action is to use these wastes as sources of food or energy in order to increase our use of renewable resources and to decrease insults to the ecosystem. Although this conclusion is axiomatic, the scientific, technological, and economic details of how this is to be accomplished have been only superficially explored.

The great diversity of microorganisms, their adaptability to various substrates and conditions, their ability to synthesize many useful nutrients, their rapid growth rate, and their ability to serve as biological filters have made it very tempting to explore the possibilities of growing microorganisms on industrial wastes for the production of single-cell protein (SCP). This paper examines the possibilities, advantages, and difficulties in the production and utilization of SCP produced by the growth of microorganisms on industrial wastes.

Three major sources of industrial wastes are, in order of magnitude, chemical-related industries, paper and allied products, and food industries (American Chemical Society, 1969). Together, these industries account for 90 percent of the biological oxygen demand produced by the industry before waste treatment and over twice that of domestic sewage (Table 1). These industries also produce over 60 percent of the settleable and suspended solids produced by industry. Primary metal production emits a large amount of solids, but the low biological oxygen demand (BOD) indicates a low organic matter content. Of these three major sources of solid wastes, those of the food and paper industries consist largely of natural products not suitable for the purposes of the industry. Chemical industry wastes are chemicals that cannot be economically recovered or for which there is little industrial use, but contain oxidizable carbon and hydrogen as evidenced by the high BOD oxygen of these effluents.

The type of fermentation and the economic and engineering aspects of SCP production depend largely on the kinds of waste and the major thrust of the recovery scheme. The food industry discards large amounts of fibrous material in the separation of highly available nutrients for human or animal use. The more available of these wastes, such as mill feeds and fermentation grains, are profitably used as animal feeds. Even high-fiber wastes, such as cottonseed hulls and cannery wastes, are usually fed to livestock. In a sense, SCP is being produced *in vivo* from these wastes by passing them through the rumen. The production of SCP from more refractory wastes, such as peanut or rice hulls, must involve a decision as to the possible recovery of energy from these sources through fermentation as compared to alternative uses of the material as litter, insulation, etc., or incineration

TABLE 1 Industrial and Domestic Wastes in the United States for 1963[a]

Source of Wastes	BOD (Million lb)	Suspended and Settleable Solids (Million lb)
Food and related	4,300	6,600
Paper, wood, and related	5,900	3,000
Chemical and related	9,700	1,900
Primary metals	480	4,700
Other industry	2,070	1,510
Domestic wastes	7,300	8,800

[a]From American Chemical Society (1969).

for power production. Production of SCP from mesquite and vinasse has been reported (Yang *et al.*, 1972; Achremowicz and Bujak, 1968). This system produces SCP, using cellulose and other polysaccharides as substrates. However, the large amount of lignin usually associated with cellulosic substrates constitutes a vast separation and disposal problem that places the economics of such systems in serious doubt.

Food industries also produce a considerable tonnage of dry matter dissolved or suspended in effluent waters. This material is soluble and often highly useful nutritionally but is in such dilute solution that recovery that involves dewatering is not economical. If production of SCP from this type of waste is to be preferable to the isolation of the original material, the process must either upgrade the product or make a product that is more easily separable. A typical example of this type of waste is whey, which is a relatively high-quality food used in some food manufacturing and a useful ingredient in livestock feeds. However, the costs of separating whey solids makes the product, except for a small volume for specialty uses, too expensive in relation to its nutrient content. On the other hand, the surplus cannot simply be discarded because of its pollution potential. Unless the manufacturer is willing to dispose of this material by lagooning or other destructive disposal methods, he may either dry the whey and sell it at a loss, charging the loss against his pollution control costs, or else he might use it to produce SCP and hope to make the income cover the expenses of the fermentation process. His choice will depend on which system results in the smallest loss. The problem is intensified by the lack of centralization in cheese manufacturing and in most food-related industries generally. The economic complexities are enough to challenge a trained economist, let alone a layman scientist. It is, however, the duty of the animal scientists to supply information on the relative nutritional value of dried whey versus SCP as a necessary input to decision making.

The paper and wood industries, in contrast to the food industry, recover most of the fibrous material and so their most abundant wastes are these dissolved and suspended solids. SCP production from this type of waste is usually incidental to the reduction in BOD obtained by microbial growth in the effluents and little attention is paid to yield of SCP per unit of waste dry matter. Indeed, in many situations, the amount of dry matter removed per unit of SCP recovered is expected to be high.

Examples of this type of SCP production are microorganisms grown on paper and wood industry effluents and activated sewage sludge. We have worked with a product that is produced by the fermentation of soluble and suspended solids in the effluents of a pressed wood products operation. The primary effort in this process is to reduce the BOD of the effluents that are produced at over 25,000 liters/hr to meet pollution control specifications. During this process, however, there is produced about 20,000 kg/day of a recoverable SCP product. The process must be maintained in an aerobic condition, and most of the solid wastes are removed by fermentation. SCP production is not the primary objective but may contribute to reducing the cost of the operation and to the recovery of some nutrients.

Effluents from chemically related industries contain relatively less suspended material but more specific and defined dissolved compounds. Finn (1970) has described many common chemical industry waste compounds (organic acids, alcohols, aldehydes and ketones, chlorinated organics, and aromatic derivatives) and has proposed methods for their utilization. Considerations of economics and engineering largely determine whether maximum yield or maximum degradation of substrate is to be sought. Because of metal catalysts and other reactive additions, the problems of contamination and toxicity are more likely to be encountered with SCP produced from chemical wastes than with effluents from food and paper industries.

SCP CULTURE

The methods of producing SCP vary greatly, ranging from simple incubation to very elaborate systems of production with complete environmental and nutrient control. Each process varies depending upon the source of substrate and the characteristics of the organisms. Several factors, however, are characteristic of many of the processes. The majority are aerobic, so that some means of aeration must be utilized (Peppler, 1967; Arnold and Steel, 1958). Oxygen injection, agitation, air pumping, or large shallow vessels are used to achieve aerobiosis. Most processes have some measure of temperature control either to maintain the culture at an optimum temperature or to prevent the accumulation of heat by the fermentation mixture. The utilization of energy in industrial wastes usually requires an added source of nitrogen so that some form of this nutrient is added in many processes (Golueke *et al.,* 1969; Gasser and Jeris, 1968). The regulation of pH and electrolyte supply is also critical in most systems. When algae are grown on wastes, the light supply, intensity, and penetration must all be controlled (Oswald and Golueke, 1968; Oswald, 1969).

Some of the above-mentioned factors are optimized simply by the design of the system while others require considerable engineering and power inputs to be achieved. Controls of the fermentation process must be justified in terms of increased yield or a higher quality product.

CELL HARVEST

A major difficulty in the use of SCP is harvesting the cells. Currently used methods include dehydration, filtration, centrifugation, and flocculation. Algae are generally more easily separable than yeasts and yeasts more than bacteria (Oswald and Golueke, 1968; Wang, 1968; Oswald, 1969; and Johnson, 1967). The ease of separation of the larger cells would lead one to believe that these organisms should be the ones of choice. Unfortunately, the bacteria are frequently able to utilize a wider variety of substrates with more efficiency and, therefore, must be considered for certain fermentations (Finn, 1970). This is especially true where refractory substrates such as cellulose are being used (Yang *et al.*, 1972).

The development of technological innovation for achieving cell harvest is critical to the whole concept of SCP production. If more efficient methods for product separation are not devised, it is questionable whether SCP production can be justified economically or from the standpoint of bioenergetics. If, on the other hand, methods can be devised that will allow inexpensive, efficient separation of SCP products, then SCP production offers an excellent method for the separation of otherwise expensively available or unavailable nutrients as, for example, those in whey.

Flocculation and some forms of filtration appear to hold more potential than dehydration or centrifugation since the latter two methods have a relatively unalterable theoretical energy input for separation. Novel methods such as reverse osmosis or electrostatic classification may also be considered.

UTILIZATION OF SCP

There is a great volume of literature concerning the utilization of SCP from different sources. Many excellent reviews have been prepared (Snyder, 1970; Mateles and Tannenbaum, 1968). The most generally mentioned problems in the use of SCP are palatability, digestibility, nucleic acid content, toxic or harmful residues, and a relatively low content of sulfur-containing amino acids. Fortunately, not all SCP products have all of these problems, but each possibility must be considered in the evaluation of SCP as animal or human food.

Palatability

Unprocessed microbial cells are usually not palatable. Yeast has a characteristic bitter flavor, and algae and bacteria, too, have flavors that, if not unpleasant, are at least unusual and therefore less acceptable than conventional foods (Lee *et al.*, 1967; Waslein *et al.*, 1969). The palatability varies with species. Humans are much more sensitive than most domestic and laboratory animals, since food has such strong psychological and social meaning to the human (Waslein *et al.*, 1969). We have tested an SCP product that was highly unacceptable to rats. The inclusion of six artificial flavors singly and in combination did not improve the acceptability of this product. Even supplementation with methionine was ineffective in improving intake of the material (R. K. Chapman, unpublished data). However, when the material was treated with 5 percent hemicellulose extract and 2 percent fat to reduce dustiness, sheep consumed as much as 250 g/day. This material fed to rats appeared to be more acceptable than the nontreated product. This work indicates that much more must be done in the area of acceptability, however, and species differences must be recognized. Some of these problems are easily corrected, while others are not. Yeast can be debittered and several methods have been proposed to obtain protein concentrates from microorganisms (Hedenskog *et al.*, 1970). The resulting products are usually bland and light colored and so are compatable with mixed foods or feeds. Synder (1970) stated, "The most promising immediate answer to the palatability problem is to use single-cell protein as a supplement to conventional foods at levels that are undetected or not objectionable."

Digestibility

There is a great variation in the digestibility of SCP products. In general, all products must be treated to kill the cells, or digestibility is very poor. This is especially true of yeasts and algae (Miller, 1968;

Lee *et al.*, 1967). In addition to killing cells, many products respond to further treatment. Algae seem to have the most response to heat treatment. Digestibility is often doubled by some form of cooking (Tannenbaum and Miller, 1967). On the other hand, there is little value beyond cell death in treating yeast. Apparently, the poor digestibility of yeast is not due to cell integrity, as in algae, but in a reduced availability of a chemical nature of cell wall constituents (Tannenbaum, 1968). It is possible to separate the highly digestible cell contents from the less digestible cell walls, but this represents isolation of the digestible fraction and not improvement of the digestibility of the cell wall fraction. Little work has been done with bacteria, but they appear to have properties similar to yeast. Reduced digestibility of bacterial nitrogen is often observed because of a substantial content of diaminopimelic acid or muramic acid that are not absorbed (Mason and White, 1971; Mason and Milne, 1971). Other cell wall constituents of bacteria may seriously reduce the digestibility of the cells.

Nucleic Acid Content

The importance of the nucleic acid content of SCP is controversial. Microorganisms usually contain significant amounts of nucleic acids, but again the variation is great (Table 2). It has been suggested that a high nucleic acid intake might result in elevated blood uric acid levels and aggravate related synovial pathology or secretory problems (Carter and Phillips, 1944). Waslein *et al.* (1968) were unable to demonstrate elevations of blood urate when pure nucleic acids equivalent to that contained in 10–20 g of yeast were fed to humans. In subse-

TABLE 2 Nucleic Acid Content of Foods and Microorganisms[a]

Material	Protein (% Dry Wt.)	Nucleic Acid (% Dry Wt.)	Nucleic Acid (% Protein)
Liver	65.6	2.6	3.9
Sardine	64.2	1.4	2.2
Fish roe	70.3	4.1	5.7
S. aureus	75.5	11.6	15.4
B. anthracis	58.1	4.4	7.5
E. coli	78.5	12.8	16.3

[a]From Miller (1968).

quent experiments, however, Waslein *et al.*(1970) used 25 and 50 g of yeast or algae and noted greatly increased uric acid excretion and significant increases in blood urate levels. In view of these results, and since purine and pyrimidine metabolism is so central to most metabolic processes, it would seem to be extremely unwise to introduce into the human diet a significant increment of nucleic acid without detailed, long-term studies on the impact of this dietary innovation. It appears that such studies are the most logical next step in the science and technology required to make SCP a possible source of human food. This is a question that must be answered satisfactorily and thoroughly. With current consumer hysteria, it will also be necessary to demonstrate safety for human consumption of the flesh of animals fed high levels of nucleic acids. A large field awaits the prepared investigator.

Toxicities

The possibility of toxicity must always be dealt with. There are two possible sources of harmful compounds in SCP. The first arises from the substrate as it comes from the industrial process, occurs as a result of concentration of dilute substances during recycling, or is selectively taken up by the organism. Those toxins specifically produced in microbial cells as regular or secondary metabolites are the second source. Consistently occurring toxins can be eliminated, but secondary metabolites or toxins resulting from mutations or environmental alteration are very difficult to control or detect. The whole subject of toxicology is too large to be covered here, but constant vigilance will be necessary.

Protein Quality

The value of SCP in human food and monogastric diets naturally depends on the quality of the protein. This has been extensively studied and most data for most products indicate a relative deficiency of sulfur-containing amino acids (Table 3). Other amino acids that may be marginal are lysine and isoleucine (Miller, 1968; Waldroup, 1972). This is a relative deficiency in many instances and the amino acid spectrum in SCP is usually superior to that in staple cereals. It is, however, usually inferior to more commonly used protein supplements (Miller, 1968). Many SCP products are equivalent to casein or soybean meal when supplemented with methionine

TABLE 3 The Effects of Supplementation of SCP[a]

	Biological Value for Rat (%)
Torula	32–48
Torula and methionine	88

[a]From Bressani (1968).

(Table 3). Fortunately, methionine is the least expensive and most commercially available amino acid. It is fortunate that most animals use the D and L forms of methionine equally well. It is therefore doubtful that the somewhat reduced protein quality of SCP will seriously affect its usefulness if all other objections are overcome.

In considering the nutritional value of SCP and its possible alteration by processing or supplementation, it must be kept in mind that SCP can be justified as food or feed only if it is cheap—economically and biologically. Processing costs must be kept to a minimum both in terms of dollars and resources. The growth of SCP is justified in terms of its ability to conserve energy resources. Therefore, the energy input during processing should not exceed, or even approach, the energy yield.

It is obvious that the need for SCP production for resource conservation and pollution control is urgent and that the production of microorganisms is possible under a wide variety of circumstances and with a wide variety of industrial wastes. This, however, represents

TABLE 4 Effect of Time of Hydrolysis on Amino Acid Analysis[a]

Amino Acid	10 hr Value × 100 Maximum Value (%)	Hydrolysis Time for Maximum Value (hr)
Serine	88	8
Proline	76	6
Cystine	12	1½
Methionine	55	6
Valine	89	24
Lysine	78	18
Histidine	84	24

[a]Sample 878. Optimum time 10 hr.

only a small fraction of the accomplishment necessary to bring to fruition the latent advantages of such recycling of energy resources. It appears that it is necessary now for us to take a fresh and unified look at research in this field to make our efforts more efficient and communications easier.

As animal scientists, we are perhaps most interested in the measurement of the biological impact of SCP as a diet ingredient. As one studies the literature in this area, one is struck by the fragmented approach to the study of the nutritive value of SCP. Many experiments yield questionable results, and there are many unexpected pitfalls as attempts are made to obtain a valid measurement of the nutritional properties of SCP.

CHEMICAL EVALUATION

With modern analytical procedures available, it is no longer satisfactory to describe the composition of a protein as in this example from a popular article: "The protein can be dried to a powder that is odorless, slightly acidic, and white to cream in color. It contains measurable amounts of the 10 essential amino acids."

The introduction of chromatographic methods of amino acid analysis has made it relatively simple to obtain values for amino acids in protein hydrolyzates. SCP products are, in the very early stages of development, routinely analyzed for amino acid content. These data give us a distinct advantage in assessing the prospective value of new products. Caution must be exercised, however, in interpretation of amino acid analysis figures. It is a protein hydrolyzate and not a protein that is separated and measured during amino acid analysis. We understand very little about hydrolysis loss in conventional proteins and even less about alterations during hydrolysis of SCP.

The usual procedure for hydrolysis in amino acid analysis is autoclaving in evacuated tubes in 6N HCl for 6 hr. We ran a series of hydrolyzates of two samples of SCP autoclaved at 0, 1½, 3, 6, 8, 10, 18, or 24 hr. Of course, under conditions of acid hydrolysis, tryptophan is destroyed, and, therefore, values for this acid were not determined. For one sample, the hydrolysis time that yielded maximum values for the most amino acids was 10 hr (Table 4), while for the other it was 6 hr (Table 5). However, several amino acids did not yield maximum values at the same time. This was especially true of cystine and methionine. These data indicate that no one hydrolysis time will give

TABLE 5 Effect of Time of Hydrolysis on Amino Acid Analysis[a]

Amino Acid	6 hr Value × 100 Maximum Value (%)	Hydrolysis Time for Maximum Value (hr)
Serine	85	18
Isoleucine	86	18
Histidine	86	18
Argine	91	10
Diaminopimelic	0 vs. 1.8	8

[a]Sample 973. Optimum time 6 hr.

the true picture of the amino acid content of a protein. Some amino acids are released early and destroyed by prolonged hydrolysis, while others are not freed by hydrolysis until some time has elapsed (Table 6). Sample preparation remains the largest problem in amino acid analysis.

Great confusion as to the methods of reporting amino acid analysis exists in the literature and makes it very difficult to compare data. It is most fundamental that the moisture status or moisture level of the samples be given. It may be necessary to determine nitrogen and amino acids on fresh samples to prevent nitrogen loss or amino acid alteration during drying, but the moisture content of the sample *must* be presented if this is the case.

The practice of reporting amino acids per 16 g nitrogen has been widely used to reduce variation due to protein level in a sample. How-

TABLE 6 Effect of Time of Hydrolysis on Amino Acid Analysis[a]

Amino Acid	$\frac{\text{18 hr Value}}{\text{6 hr Value}} \times 100$
Glutamic	86
Proline	77
Glycine	89
Alanine	80
Cystine	48
Methionine	36
Tyrosine	71
Phenylalanine	76

[a]Sample 973. Time 18 hr.

ever, when SCP samples vary from 2 to 12 percent in nucleic acids and perhaps even more than that in other nonprotein nitrogen, comparisons are still not valid. Of immediate importance is a method for the accurate determination of true protein that will consistently work for SCP. There are strengths and weaknesses in the various methods of expressing amino acid content of SCP, and it is not the purpose of this paper to discuss their relative merits. In the interests of communication, however, it would seem wise to insist upon the inclusion of data necessary to convert one method to another so that valid comparisons can be made.

BIOLOGICAL EVALUATION

Perhaps the least satisfactory design used is a test of the partial substitution equivalence of SCP for a standard protein source. Without a negative control, such experiments may only indicate how much dilution the original diet could tolerate. The fallacy of such a design is illustrated by the following experiment on the substitution of urea or soybean meal in rations for fattening steers. Steers without any protein supplement performed as well as those receiving either supplement during the last 56 days of the trial (Table 7). Without this negative control, it would have been falsely assumed from the data that urea and soybean meal were equivalent sources of nitrogen for steers under these conditions when, in fact, nothing could be said for any of them since nitrogen was adequate in the basal diet.

Even with negative controls, results of this type of experiment must be interpreted with great caution. It is very possible that poor performance may be the result of toxic, indigestible, or unpalatable factors in the SCP and not of basic nutritional inadequacy. Frequently, such factors may be corrected by processing and a highly

TABLE 7 Influence of Protein Supplement on Cattle Gain

	Cattle Gains (kg/day)
Soybean meal	1.14
Urea	1.12
No supplement	1.15

nutritious product results. This confounding of reasons for poor response to substitution of SCP for other protein is a pitfall that must be avoided. Substitution trials are easy to run and perhaps may serve as screens for more intensive studies but can never be considered conclusive. In addition to being unsatisfactory from a scientific standpoint, the reasoning that, "It was possible to substitute SCP for 25 percent of the protein supplement without reducing performance," is also poor politics, since it implies that the investigator assumed that the material was inferior or harmful and was looking for an acceptable level under which the product might be tolerated.

The use of PER avoids the problem of the negative control if the basal diet is sufficiently low in protein but will still confound nutritional adequacy with toxic and aesthetic factors of the feed. Palatability differences can be partially controlled using controlled feed intake. The test animal to be used is critical, since marked differences between species have been observed.

Net protein utilization (NPU) gives a more accurate picture of the amino acid adequacy of SCP but this necessitates collection of feces and urine and considerably more expense. The validity of NPU is also ambiguous. It must be conducted at a nitrogen intake below the animals' requirements and with adequate energy. More importantly, unless the SCP being tested is the only source of protein, the technique measures the supplementation potential of the SCP with regard to the basal protein. This would vary widely with the protein in the basal ration. This is not entirely bad, since SCP is meant to be a supplemental protein (Table 8). However, definitive statements with regard to the value of SCP would have to be made specifying the source and level of basal protein. Such data will be needed before a SCP product is released for human or animal use.

TABLE 8 SCP as Protein Supplement[a,b] (Calculated)

Material	Limiting Amino Acid	Limiting Amino Acid as % Essential Amino Acid		Protein Score	
		Before	After	Before	After
Wheat	Lysine	5.9	7.4	47	59
Corn	Lysine	5.4	6.9	43	55
Corn	Tryptophan	1.4	2.9	45	94
Peanut	Sulfur amino acids	4.5	6.1	42	60

[a] From Miller (1968).
[b] Using *B. megaterium* as 20 percent of protein intake.

All of the mentioned tests are short term. The nature of SCP requires that it be tested thoroughly throughout the life cycle of one or more species. This will not only help unmask unknown deficiencies or toxicities but will also help determine the long-term effects of such known ingredients as nucleic acids, unusual fatty acids, or atypical amino acids that are poorly utilized and must be disposed of by the body. The extended effects of these materials on circulation, skeletal growth and excretory processes, among others, must be ascertained.

We believe it can be seen that there is no one biological test for a SCP product that will answer all questions. The dangers inherent in premature introduction of SCP products are such that every conceivable safeguard should be applied before these products are made generally available.

INTERDISCIPLINARY ASPECTS

In this paper, we have not attempted to discuss in detail the findings of an army of investigators whose results are widely published and adequately reviewed, but rather to try to bring into focus for the purpose of discussion some ideas that will make the contributions of animal scientists to this very important field of endeavor more efficient and productive.

It is very difficult, in considering the whole subject of SCP, to avoid lengthy involvement in economics, engineering, and related fields. Indeed, the biological aspects are sometimes only peripheral to these other considerations. It is, therefore, essential that a multidisciplinary team approach to the subject be sought. Absence of such consultation makes any work in this area suspect indeed. On the other hand, economic and engineering aspects must be guided by accurate knowledge of the nutritional and biological potential of the resulting product.

In order for development of sound SCP systems, an order of economics above that of dollars input and output must be applied. The energy economy of the ecosystem must be known and energy expenditure justified. In this connection, it is very difficult to find energy input and output figures for various SCP schemes. In most cases, especially where SCP production is incidental to pollution control, there are no data available that indicate the dry matter inputs into the fermentation compared to the SCP produced.

In addition to raw material input and product output, the energy cost of processing and distribution must be considered. Dehydration, centrifugation, and other separation methods require a large input of energy—often greater than the energy value of the product. In this case, SCP is not a renewable resource but merely a conversion of fossil energy to edible protein and energy.

The importance of SCP production has been established and the potential technical feasibility has been demonstrated. There now remains the greater task of discovering those facts that will allow us to develop a system of SCP production justifiable in economic and biological terms. This is the most difficult aspect. It will require a concerted and objective approach by large teams of experts, not the least important of which will be animal scientists. Livestock are necessarily an intermediate link in the use of SCP, but it is very possible that in the long run SCP will be justifiable only as human food. It is our challenging responsibility to see that such food is as nourishing and safe as possible.

REFERENCES

Achremowicz, B., and S. Bujak. 1968. Experimental selection of yeast-like organisms for the production of feed protein on vinasse. Ann. Univ. Mariae Curie-Sklodowska, Sec. E 23:391–401 (Poland). [Chem. Abstr. 73:191 (1970).]

American Chemical Society. 1969. Cleaning our environment. The chemical basis for action. American Chemical Society, Washington, D.C. 249 p.

Arnold, B. H., and R. Steel. 1958. Oxygen supply and demand in aerobic fermentations, p. 151–181. *In* R. Steel [ed.], Biochemical engineering. Unit processes fermentation. Heywood and Co., London, 328 p.

Bressani, R. 1968. The use of yeast in human foods, p. 79–80. *In* R. I. Mateles and S. R. Tannenbaum [eds.], Single-cell protein. M.I.T. Press, Cambridge, Mass.

Carter, H. E., and G. E. Phillips. 1944. The nutritive value of yeast proteins. Fed. Proc. 3:123–128.

Finn, R. K. 1970. Microbial cells from wastes as a feed supplement. Proc.1970 Cornell Nutr. Conf., Ithaca, N.Y. 115 p.

Gasser, R., and J. S. Jeris. 1968. Comparison of various nitrogen sources in anaerobic treatment. Water Pollut. Control Fed. J. 39:R91–R100.

Golueke, C. G., W. J. Oswald, and H. K. Gee. 1969. Effect of nitrogen additives on algal yield. Water Pollut. Control Fed. J. 41:823–834.

Hedenskog, G., H. Mogren, and L. Enego. 1970. A method for obtaining protein concentrates from microorganisms. Biotechnol. Bioeng. 12(6):947–959.

Johnson, M. J. 1967. Growth of microbial cells on hydrocarbons. Science 155:1515–1519.

Lee, S. K., H. M. Fox, C. Kies, and R. Dam. 1967. The supplementary value of algae protein in human diets. J. Nutr. 92:281–285.

Mason, C. V., and G. Milne. 1971. The digestion of bacterial mucopeptide constituents in the sheep. 2. The digestion of muramic acid. J. Agric. Sci. 77:99–101.

Mason, C. V., and F. White. 1971. The digestion of bacterial mucopeptide constituents in the sheep. 1. The metabolism of 2, 6-diaminopimelic acid. J. Agric. Sci. 77:91–98.

Mateles, R. I., and S. R. Tannenbaum [eds.]. 1968. Single-cell protein. M.I.T. Press, Cambridge, Mass. 480 p.

Miller, S. A. 1968. Nutritional factors in single-cell protein, p. 79–89. *In* R. I. Mateles and S. R. Tannenbaum [eds.], Single-cell protein. M.I.T. Press, Cambridge, Mass.

Oswald, W. J. 1969. Current status of microalgae from wastes. Chem. Eng. Prog. Symp. Ser. 65(93):87–92.

Oswald, W. J., and C. G. Golueke. 1968. Large-scale production of algae, p. 271–305. *In* R. I. Mateles and S. R. Tannenbaum [eds.], Single-cell protein. M.I.T. Press, Cambridge, Mass.

Peppler, H. J. 1967. Yeast technology, p. 145. *In* Microbial technology. Reinhold, N.Y.

Snyder, H. E. 1970. Microbial sources of protein. Adv. Food Res. 18:85–140.

Tannenbaum, S. R. 1968. Factors in the processing of single-cell protein, p. 343–352. *In* R. I. Mateles and S. R. Tannenbaum [eds.], Single-cell protein. M.I.T. Press, Cambridge, Mass.

Tannenbaum, S. R., and S. A. Miller. 1967. Effect of cell fragmentation on nutritive value of *Bacillus megaterium* protein. Nature 214:1261–1262.

Waldroup, P. W. 1972. The future of petroleum derived protein sources for livestock. Proc. Distill. Feed Res. Conf. 27:34.

Wang, D. I. C. 1968. Cell recovery, p. 217–228. *In* R. I. Mateles and S. R. Tannenbaum [eds.], Single-cell protein. M.I.T. Press, Cambridge, Mass.

Waslein, C. I., D. H. Callaway, and S. Margen. 1968. Uric acid production of men fed graded amounts of egg protein and yeast nucleic acid. Am. J. Clin. Nutr. 21:892–897.

Waslein, C. I., D. H. Calloway, and S. Margen. 1969. Human intolerance to bacteria as food. Nature 221:84–85.

Waslein, C. I., D. H. Calloway, S. Margen, and F. Costa. 1970. Uric acid levels in men fed algae and yeast as protein sources. J. Food Sci. 35:294–298.

Yang, S. P., H. H. Yang, D. W. Thayer, and A. B. Key. 1972. Nutritional value of the microbial protein produced from cellulose waste [abstract]. Fed. Proc. 31:695.

L. W. Smith

RECYCLING ANIMAL WASTES AS PROTEIN SOURCES

Animal nutritionists are aware of the importance of animals (Cutherbertson, 1969), and especially ruminants (Reid, 1970) in meeting the world's expanding protein needs. Most are also aware of the environmental quality degradation often associated with excrement from intensified confinement systems for livestock production and the evolving regulations to ensure environmental quality. The latter development offers justification and opportunity to appraise the feasibility of shunting the classical land–plant–animal recycle system. Success will depend upon favorable production economics and demonstrated high-quality and safe animal products. Animal recycle systems for utilizing undigested feed nutrients and by-products of digestion and metabolism could place animals in an improved competitive position for supplying an even greater part of the world's future protein needs.

Various aspects of nitrogen, amino acid, and protein requirements, digestibility, and metabolism of farm animals have been recently reviewed (McGilliard, 1972; Allison, 1969; Waldo, 1968; Scott *et al.*, 1969; Becker *et al.*, 1963). These reviews provide general background information. The purpose of this review is to discuss the use of animal waste as a protein source for various classes of farm animals as related to the diversity of nitrogen compounds in animal waste and to discuss some animal recycling systems for efficient utilization.

QUANTITY OF RAW MANURE AND NITROGEN PRODUCED IN THE UNITED STATES

Taiganides and Stroshine (1971) estimated, based on 1969 statistics, that animal wastes were produced at the rate of 3,118 million kg/day on a wet basis in the United States (Table 1). They indicate their estimate was 32 percent lower than figures reported by Train *et al.* (1970).

Their estimate of the amount of nitrogen in manure excreted by the major species of farm livestock is used as the basis to further categorize and inventory animal waste nitrogen into the various chemical forms.

It should be understood that only a portion (unknown at this time) of the total estimated nitrogen in animal waste is economically recoverable as a potential protein source for animals. Under the large confinement systems (caged layer operations and beef feedlots), collection is a minor problem, and an incentive exists for considering animal waste as a protein source. The economics of collection from diffuse animal systems, however, would be prohibitive with current technology and even undesirable, since animals under these types of management are themselves admirable spreaders when land is not a restraint.

Partitioning of total nitrogen into fecal nitrogen and urinary nitrogen provides a means to gain further information on the chemical form and, at least in some cases, information on the expected utilization of nitrogen in animal waste as a protein source for animals.

TABLE 1 Quantities of Raw Manure and Nitrogen in Manure by Major Species of Livestock for 1969[a]

	Millions kg/day					
	Beef Cattle	Dairy Cattle	Sheep	Swine	Poultry	Total
Raw manure	1,491	1,165	52	184	22.6	3,118
Total nitrogen	20.0	3.4	0.6	1.4	3.4	28.8

[a]From Taiganides and Stroshine (1971).

Benedict and Ritzman (1923) reported data from which one can calculate that about 50 percent of excreted nitrogen appeared in feces of a fattening steer. Current data (D. A. Dinius, personal communication, 1972) confirm that observation. Our data (L. W. Smith and C. C. Calvert, unpublished data) show a similar nitrogen distribution between feces and urine from sheep as from beef cattle. Conrad *et al.* (1960) summarized nitrogen balance data from nine sources as well as results from their experiments with lactating dairy cows. In these trials, 47–70 percent of the excreted nitrogen appeared in feces, the mean being 60 percent. Morrison (1957) reported that, for swine, about 67 percent of the total nitrogen excreted was in feces. This value is in wide disagreement with current observations of about 33 percent of total nitrogen in feces from swine (R. J. Davey, personal communication, 1972). Calculations in this paper are based upon 33 percent nitrogen in feces and 67 percent in urine for swine. Poultry excrete about 25 percent of nitrogen in feces (Breon, 1939; and Bolton, 1954).

Distributions of nitrogen in feces and urine used in this presentation are summarized in Table 2. These values were used as a first step in estimating the quantities of nitrogen in various nitrogenous compounds in animal waste. It should be understood that not all animals, even within species, excrete fecal and urinary nitrogen in these fixed proportions. The distribution of nitrogen in feces and urine of animals is a function of several nutritional and physiological factors. These estimates, however, should be representative, so that the major nitrogenous compounds are reasonably inventoried.

Nitrogenous compositions of urine from dairy cattle (Morris and Ray, 1939b; Hart *et al.*, 1911), sheep (Morris and Ray, 1939a), swine (Stekol, 1936), and poultry (O'Dell *et al.*, 1960) were used to esti-

TABLE 2 Distribution of Nitrogen in Feces and Urine for the Major Species of Farm Animals

	% of Total N	
Species	Feces	Urine
Beef cattle	50	50
Dairy cattle	60	40
Sheep	50	50
Swine	33	67
Poultry	25	75

mate the amounts of urinary nitrogen as amino acid, urea, ammonia, uric acid, and as other forms.

Recognizing that the nitrogen distribution in urine is variable and influenced by ration and analytical procedures, several adjustments were made in some of the data that showed especially high ammonia nitrogen content. For ruminants, a common distribution of nitrogen (urea nitrogen, 71 percent; ammonia nitrogen, 1.0 percent; amino acid nitrogen, 8.3 percent; and in other forms, 19.7 percent) was derived for use in this paper. Calculations showed that swine urine (Stekol, 1936) contained about 75 percent of nitrogen in the form of urea. No additional analytical data were available to develop a more complete and realistic partitioning of swine urinary nitrogen. Therefore, the remaining 25 percent was assumed to be other nitrogen. The analytical distribution of nitrogen in poultry urine was used as presented by O'Dell *et al.* (1960) for layers fed a practical ration.

Amino acid data for beef cattle feces (Anthony, 1971), sheep feces (Loosli *et al.*, 1949), dried swine feces (Orr, 1971), and dehydrated poultry waste (Flegal and Zindel, 1970) were recalculated on the basis of g amino acid/16 g fecal nitrogen (Table 3).

TABLE 3 Amino Acid Content of Farm Animal Feces

	Amino Acid g/16 g N			
Amino Acids	Beef Cattle	Sheep	Swine	Poultry[a]
Arginine	1.1	3.4	3.1	4.2
Histidine	0.7	1.3	1.8	1.8
Isoleucine	1.3	3.7	4.8	4.5
Leucine	3.8	4.4	7.2	7.3
Methionine	2.9	5.1	5.1	4.4
Phenylalanine	0.6	1.5	2.7	0.8
Threonine	0.0	3.4	4.0	4.0
Tryptophan	1.8	4.8	3.7	4.5
Valine	–	0.9	–	–
Aspartic acid	2.3	4.9	4.8	5.7
Serine	4.3	–	6.3	9.7
Glutamic acid	1.5	–	2.7	4.7
Proline	3.8	–	15.6	14.0
Glycine	1.8	–	4.2	4.9
Alanine	2.7	–	7.0	7.5
Cystine	4.0	–	5.3	9.7
Tyrosine	0.2	–	3.0	2.3
Total amino acids (g)	32.5	33.4	81.9	100.0

[a] g amino acid/16 g true protein nitrogen.

The quantities of animal waste nitrogen in the major nitrogen excretory products are presented in Table 4. These values are presented in millions kg/day and include both fecal and urinary nitrogen.

The remaining nitrogen not accounted for in amino acid in feces was included as other nitrogen. Perhaps values calculated underestimate the amount of amino acid nitrogen in beef cattle waste, since higher values were reported by Anthony (1969). Dairy cattle feces were assumed to be of the same amino acid content as beef cattle feces since no comparable data for dairy cattle feces were available. This assumption may introduce some error in the estimate since dairy cattle feces are generally higher in fiber content and lower in fecal nitrogen content and, therefore, may not necessarily be similar to beef cattle feces in amino acid content. These differences would be expected due to differences in forage: concentrate ratios of rations.

The amino acid data (Flegal and Zindel, 1970) for poultry feces were not adjusted even though presented as g amino acid/100 g true protein. Bolton (1954) suggests that approximately 90 percent of nitrogen in poultry feces is true protein nitrogen. Therefore, the value for daily production of amino acid in poultry waste in Table 4 may be 10 percent too high.

Other nitrogen products (in the bottom line of Table 4) includes the uric acid nitrogen for animals other than poultry, purine base nitrogen, hippuric acid nitrogen, creatinine nitrogen, creatine nitrogen, allantoin nitrogen or other nitrogenous products that were not determined. Generally speaking, this category would be expected to have little nutritional value for recycling to monogastric animals. Only 21.8 percent of the total animal waste nitrogen is identified as amino acids, whereas the remaining 78.2 percent is nonamino acid nitrogen or amino acid nitrogen not accounted for by methods in

TABLE 4 Amounts of Nitrogen in Excretory Products

	Millions kg/day						
Nitrogen Product	Beef Cattle	Dairy Cattle	Sheep	Swine	Poultry	Total	% of Total
Amino acid	4.09	0.77	0.13	0.38	0.91	6.28	21.8
Urea	7.10	0.96	0.21	0.71	0.12	9.10	31.6
Ammonia	0.10	0.01	<0.01	–	0.27	0.38	1.3
Uric acid	–	–	–	–	2.06	2.06	7.2
Other	8.71	1.66	0.26	0.31	0.04	10.98	38.1

earlier reports used to develop this summary. This leads to the conclusion that animal waste nitrogen that has undergone essentially no additional processing or conversion would be more successfully utilized by ruminants than nonruminants.

Three critical questions remain unanswered: Can farm animals consume animal waste and utilize the nitrogen? Which animals have the greatest capability for utilizing specific animal waste nitrogen? What level of productivity can be expected from livestock consuming substantial quantities of animal waste? This review will discuss some of the literature on pertinent aspects of the above questions.

USE OF WASTE NITROGEN AS A PROTEIN SOURCE IN PERSPECTIVE

Nitrogen in feces from monogastric animals (nonruminants) is expected to be more usable because digesta undergoes gastric digestion, followed by a proliferation of microorganisms in the lower gut that escape digestion. In the ruminant, digesta undergoes a microbial fermentation followed by gastric digestion. Ruminants are capable of digesting greater amounts of cellulose than monogastrics (Keys *et al.*, 1969) and utilizing nonprotein nitrogen (Virtanen, 1966) and possibly bacterial cell wall nitrogen (Hoogenaard *et al.*, 1970) not utilizable by monogastric livestock.

Amino Acid Nitrogen

Generally, only a small proportion of the total excreted amino acids are in urine. These are probably free amino acids. The remainder, greater than 90 percent, are of fecal origin. There are only limited data on the origin and distribution of nitrogen in livestock feces. It could be either as undigested dietary, metabolic, or bacterial nitrogen. Mason (1969) reported on methods for partitioning sheep fecal nitrogen relative to its origin. He concluded that 51–81 percent of the nondietary fecal nitrogen was in the form of bacterial material and suggested that most had originated in the rumen. Perhaps the fecal-nitrogen pattern of other ruminants falls within this range. Research has apparently not been conducted on the form in which nitrogen is excreted by poultry and swine. Zebrowska (1968) reported data that suggest endogenous nitrogen excretion is affected by the type of protein fed to rats.

In single-stomached animals, nitrogenous by-products of gastric digestion and undigested dietary nitrogen could be converted to bacterial nitrogen in the large intestine and cecum. Reported values suggest that cecal contents of pigs (Willingale and Briggs, 1955) and poultry (Smith, 1965) contain about the same number of bacteria (10^8–10^{10}/g) as ruminal fluid that contains about 10^{10} bacteria/ml (Hobson, 1969). An important difference is that ruminal bacteria undergo gastric digestion prior to arrival in the lower gut. Viable bacterial counts in swine and poultry feces were higher than in ruminant feces (Smith, 1965).

Since both enteric bacteria and ruminal bacteria are predominantly Gram-negative, it may be valid to speculate on the former based upon studies with the latter. True digestibility of ruminal bacterial nitrogen ranged from 55 to 82 percent when fed to rats (Johnson *et al.*, 1944; McNaught *et al.*, 1950, 1954; Reed *et al.*, 1949; Bergen *et al.*, 1968; and Mason and Palmer, 1971). Some of the wide differences in digestibility could be attributed to differences in isolation and preparation of the bacteria before feeding to the rats.

Whole cells, cell contents, or cell wall preparations of ^{14}C-labeled *Bacillus subtilis* and *Escherichia coli* were 85–97.5 percent digestible when introduced into the rumen or abomasum of sheep (Hoogenraad *et al.*, 1970). *B. subtilis* and *E. coli* were chosen as being representative of Gram-positive and Gram-negative bacteria. The cell walls were more digestible than whole cells when both were introduced in the abomasum so as to avoid ruminal fermentation. Peptidoglycan or mucopeptide, a major component of the cell wall, was shown to be resistant to digestion by trypsin (Salton and Pavlik, 1960). Hoogenraad *et al.* (1970) raised the question on the presence of bacteriolytic enzymes in sheep intestine because of the high digestibility of bacterial cell wall introduced in the abomasum. In conflict with Hoogenraad *et al.* (1970), Mason and Milne (1971) and Mason and White (1971) concluded that bacterial mucopeptide synthesized in the rumen was not digested in the small intestine, but that 2,6-diaminopimelic acid and mucopeptide in the bacterial cell wall material was extensively degraded by bacteria in the cecum and colon. Hoogenraad and Hird (1970) showed that whole ruminal bacteria were comprised of about 40 percent and bacterial walls of about 30 percent amino acids with little difference in specific amino acid distribution, except for the higher content of 2,6-diaminopimelic acid in the cell walls.

Mason and Palmer (1971) concluded that about 20 percent of

ruminal bacteria nitrogen was not absorbed by the rat and was mainly associated with the bacterial cell wall, but that 71 percent of the absorbed nitrogen was utilized for rat tissue synthesis. The work of Hoogenraad *et al.* (1970) indicates that sheep utilize bacterial cell walls as well as cell contents. Results of these studies suggest that ruminants would utilize amino acid nitrogen in bacteria and especially bacterial cell walls more fully than other livestock.

Urea Nitrogen

Based upon data in Table 4, urea nitrogen is the largest identified nitrogenous excretion product in animal waste. Virtanen (1966) and many other investigators have demonstrated the usefulness of urea in ruminant nutrition. Urea is most favorably utilized in ruminant rations high in digestible carbohydrates. Animal wastes are generally low in digestible carbohydrates, especially when recycled within the same species. Nonprotein nitrogen has not been demonstrated to be of any nutritional value to swine or poultry (Stangel *et al.*, 1963).

An important consideration bearing on the utilization of animal waste urea nitrogen is that perhaps greater than 90 percent of the 31.6 million kg produced each day is excreted in urine. If urea is to be captured for recycling, facilities must be designed appropriately for recovery. It is doubtful that any significant quantity of urea has been captured and recycled in studies discussed herein, except those by Harmon *et al.* (1971b) who has recycled oxidation ditch residues. Aerobic bacteria in oxidation ditches probably rapidly utilize urea for synthesis to microbial protoplasm so that even in this system urea is not directly recycled.

Ammonia Nitrogen

Many species of ruminal bacteria are capable of using ammonia as the only nitrogen source (Bryant and Robinson, 1963), while for others ammonia is essential for growth (Bryant and Robinson, 1962). Pilgrim *et al.* (1970) assessed the extent to which ammonia nitrogen serves as a starting point for synthesis of microbial nitrogenous compounds by continuously infusing ^{15}N as $(^{15}NH_4)_2SO_4$ into the rumen of a sheep. Minimal extents of conversion were 53–68 percent. Ammonia nitrogen has not been demonstrated to be of nutritional value to nonruminants.

Uric Acid Nitrogen

Uric acid has been shown to provide nitrogen to ruminal bacteria in laboratory studies (Belasco, 1954) when cellulose digestion was the criteria for evaluation. Washed cell suspensions of ruminal bacteria completely degraded uric acid, but at a slow rate and with the evolution of carbon dioxide, ammonia, and acetic acid (Jurtshuk *et al.*, 1955). Looper and Stallcup (1958) demonstrated that ammonia was released from chicken litter at near the same rate as from urea when incubated with ruminal inoculum.

Uric acid was evaluated as a source of dietary nitrogen in purified diets of steers (Oltjen *et al.*, 1968). Even though uric acid compared favorably and was probably superior to other NPN sources based upon nitrogen retention, differences were not statistically significant. In another study, sodium urate was superior to 25 percent sodium urate and uric acid as a dietary nitrogen source for steers based upon nitrogen retention and ruminal ammonia levels (Oltjen *et al.*, 1972).

Other Nitrogen

Animal excreta contains many nitrogenous compounds that have not been specifically quantified in reports in the literature, yet it is presumptous to assume that monogastrics necessarily utilize any of these nitrogenous excretory products. Nucleic acids are digested by enzymes in the small intestine but no nutritional significance has been established for ruminants or nonruminants. However, about 15 percent of ruminal bacteria nitrogen is in nucleic acid nitrogen (Ellis and Pfander, 1965). Ellis and Bleichner (1969) predict a low digestibility of nucleic acids in sheep cecal microorganisms.

Amide nitrogen and amidine nitrogen has supported *in vitro* growth of ruminal bacteria (Belasco, 1954; Jurtshuk *et al.*, 1955). Nucleic acids (Smith and McAllan, 1970) and 2,6-diaminopimelic acid (Mason and White, 1971) are rapidly degraded by ruminal bacteria. Peptidoglycan, a major constituent of bacterial cell walls, was digested when introduced into the rumen of sheep (Hoogenraad and Hird, 1970).

Ruminants appear to be capable of utilizing this largest category of nitrogenous compounds in animal waste while monogastrics show much less potential, or none at all, for utilizing such compounds.

FEEDING, DIGESTION, AND METABOLISM STUDIES WITH ANIMAL WASTE AS A SOURCE OF PROTEIN

Various types of animal wastes have been fed to farm livestock. The different animal wastes and recycle systems will be discussed in descending order relative to apparent nutritional feasibility.

Manure from Caged Poultry

Waste from caged poultry has been studied as a nitrogen supplement for beef cattle (El-Sabban *et al.*, 1970; Bucholtz *et al.*, 1971; Rusnak *et al.*, 1966; Bull and Reid, 1971). Nearly isonitrogenous rations of 10 percent crude protein with nitrogen provided by either soybean oil meal, autoclaved poultry manure, dried poultry manure, or urea were compared when fed to 28 301-kg Angus steers in a 139-day trial (El-Sabban *et al.*, 1970). Autoclaved poultry waste contained 35 percent and dried poultry waste 25 percent crude protein, and both products were fed at about 4.9 percent of the ration. Steers fed the dried poultry waste had the lowest average daily gain (ADG), 1.15 kg, and highest feed : gain ratio, 10.83 kg, while the urea-fed steers outgained (ADG 1.43 kg) the other groups with less feed (8.14 kg/kg gain). Steers fed the hydrolyzed poultry waste supplemented rations performed essentially equal to steers fed soybean oil meal with ADG of 1.21 versus 1.22 and feed : gain ratios of 10.0 versus 10.3. Rusnak *et al.* (1966) compared autoclaved poultry waste with soybean oil meal in rations fed to 20 Hereford steers for 89 days. Soybean oil meal and autoclaved poultry waste were both supplemented at 9.9 percent of the ration, which resulted in a crude protein content of 12.8 percent for the former and 12.9 percent for the latter ration. Soybean oil meal fed animals gained, on the average, 0.98 kg/day compared to 0.99 kg/day for the autoclaved poultry waste fed group with feed : gain ratios of 11.6 and 12.9, respectively. Bucholtz *et al.* (1971) tested dehydrated poultry waste that contained 17 percent crude protein against soybean meal or urea and combinations of 1 : 1 soybean meal to dehydrated poultry waste or 1 : 1 urea to dehydrated poultry waste in isonitrogenous rations for 314-kg yearling steers. Average daily gain for steers fed the dehydrated poultry waste containing diets gained (1.31 kg/day), significantly less ($P < 0.01$) than those fed either the urea or soybean meal diets (1.46 kg/day). Steers sorted shelled corn and corn silage from the ration, refusing to consume the dehydrated poultry waste portion. A question arises as to whether the steer groups had varying crude protein intakes that explain the differences in rate of gain. Bull and Reid (1971) determined nitrogen digestibility and balance of six growing Hereford steers fed air-dried poultry manure. A basal diet containing 5 percent crude protein was supplemented with 12, 24, and 45 percent air-dried caged layer manure. Digestibilities of dry matter, crude protein, total carbohydrate, ether extract, ash, and energy increased with the first incremental addition of air-dried poultry manure. Apparent crude protein digestibility of the air-dried poultry manure, determined by difference, was 73 percent for the lowest level and 82 percent for the highest level of addition. The percentage of absorbed nitrogen retained declined as nitrogen intake increased. This was expected since nitrogen intake exceeded calculated daily maintenance needs by factors of 2.3, 3.9, and 6.8.

Dairy cattle fed concentrate mixtures containing about 30 percent dried poultry waste appear to need an adaptation period of 7–21 days in order to achieve maximum concentrate consumption (Bull and Reid, 1971; Thomas *et al.*, 1972). Bull and Reid (1971) state that lactation persistency, milk flavor, and animal health were not adversely affected for three cows consuming 2.3–4.1 kg air-dried poultry manure/day. They suggest that air-dried poultry waste can serve as the sole source of supplemental nitrogen for cows producing up to 28 kg milk/day. Thomas *et al.* (1972) also showed that dairy cows fed dehydrated caged layer feces to provide 23 percent of total dietary nitrogen produced more milk than those fed inadequate nitrogen and produced equal amounts to those fed conventional supplements. Cow body weight changes were not influenced significantly over the 85-day feeding period (Thomas *et al.*, 1972).

Nitrogen in rations containing either autoclaved or cooked poultry wastes as sole nitrogen sources were 66 and 69 percent digestible. Nitrogen in the ration containing cooked poultry waste was not significantly different from the 74 percent nitrogen digestibility for the control soybean oil meal ration (El-Sabban *et al.*, 1970). The nitrogen digestibilities are similar to those reported by Bhattacharya and Fontenot (1965, 1966) for rations containing poultry litter. Retained nitrogen values, expressed as percent of either intake or digested, were not statistically different, but values tended to be higher for the autoclaved or cooked poultry waste ration. These retentions, expressed as percent of digested ration, are in agreement with those of Bhattacharya and Fontenot (1965) for a ration in which nitrogen was totally supplied by poultry litter. Nitrogen retention expressed as percent of intake was higher for steer diets containing uric acid nitrogen than other nonprotein nitrogen sources (Oltjen *et al.*, 1968). Nitrogen balances expressed as above were essentially identical for soybean oil meal nitrogen and urea nitrogen (Hatfield *et al.*, 1955). While it is apparent that poultry waste nitrogen may be slightly less digestible than soybean meal, nitrogen retention data suggest that absorbed poultry waste nitrogen may be more efficient for tissue deposition than other NPN sources and perhaps equal to soybean oil meal nitrogen when formulated into diets at not too high a level.

Lowman and Knight (1970) determined apparent nitrogen digestibility of dried poultry manure by direct measurement and by an extrapolative method from diets containing different levels of dried poultry manure. An apparent digestibility coefficient of 77.2 was obtained by direct measurement, whereas the extrapolated value was 78.7 percent and in very close agreement. L. W. Smith and C. C. Calvert (unpublished data) determined the true protein digestibility of dried poultry manure to be 80 percent. This value appears to be low in view of the 77–79 percent apparent nitrogen digestibility reported by Lowman and Knight (1970).

Reported values for apparent nitrogen digestibilities range between 63 (Thomas *et al.*, 1970) and 79 percent (Lowman and Knight, 1970) in sheep. This is in good agreement with values of 73–82 percent with steers (Bull and Reid, 1971).

Dehydrated poultry feces were fed to fattening sheep (Thomas *et al.*, 1972) to provide 61 or 70 percent of total dietary protein and 25 or 50 percent of feed intake. Sheep fed 25 or 50 percent dehydrated feces in rations gained 0.16 and 0.15 kg/day; this was significantly less than those fed a soybean oil meal ration (0.21 kg/day). Perhaps the relatively large differences in ADG were due to the high crude protein content, largely in the form of NPN in the dried poultry feces. The National Research Council's Committee on Animal Nutrition (1968) recommends that fattening lamb diets contain 12.0 percent crude protein, whereas the rations Thomas *et al.* (1972) fed contained 18–20 percent crude protein.

Laying hens were fed 21.6 percent crude protein unheated hen manure at either 10 or 20 percent of the diet in place of soybean meal (Quisenberry and Bradley, 1968). Hen body weight, hen-day egg production, feed efficiency, and egg size were unaffected by this isocaloric and isonitrogenous substitution.

Flegal and Zindel (1970, 1971) and Flegal and Dorn (1971) recycled manure from caged laying hens at levels up to 40 percent of the diet. Dehydrated poultry waste was substituted for corn in the rations and an attempt was made to keep calculated crude protein levels equal in all rations. Incorporation of up to 20 percent dehydrated poultry waste had no effect on egg production or feed efficiency of hens to produce eggs (Flegal and Zindel, 1971). Hodgetts (1971) also reported that 20 percent dried hen manure in the diets of hens had no adverse effect on egg production or feed efficiency for egg production. Feed efficiency for egg production was increased slightly by addition of animal fat to hen diets (Flegal and Zindel, 1971). Decline in productive efficiency on diets containing above 20 percent dried waste was not prevented by additions of lard (Pisone and Begin, 1971). These results suggest that protein in dried hen manure is less limiting for egg production than its energy content.

Feeding 0–20 percent dehydrated poultry manure in diets for growing chicks has resulted in no significant differences in body weight, but feed efficiencies were inversely related to the amount of dehydrated poultry manure (Flegal and Zindel, 1970, 1971). Data of Calvert *et al.* (1971) show a marked depression in growth rate of chicks fed diets containing 22 percent of two types of hen manure instead of soybean meal. In fact they suggested that dehydrated hen manure was only slightly better than cellulose in diets for growing chicks. Wehunt *et al.* (1960) reported that about one half of the crude protein of hen manure and one third of the crude protein of broiler manure used in their studies was true protein. Growth rates of chicks were improved when these manures were added to diets suboptimal in protein, but feed efficiency was not improved. Their calculations of g protein/g gain showed that manure-true protein was used approximately equal to that of soybean meal or a casein–gelatin mixture.

Fly pupae meal and fly meal produced from flies reared on hen manure was shown to be equal to soybean meal for growing chicks based upon similar growth rates and feed-to-gain ratios (Calvert *et al.*, 1971).

Beef Feedlot Manure

Feedlot manure has been recycled as a feed for beef cattle (Anthony and Nix, 1962; Anthony, 1966, 1969, 1970, 1971). Washed feedlot manure mixed with concentrate or concentrate and corn silage (Anthony and Nix, 1962; Anthony, 1970), autoclaved or untreated mixed with concentrate (Anthony, 1970), and fresh untreated mixed with Bermuda hay and then ensiled (Anthony, 1969, 1971), have been compared as feeds for finishing beef cattle. Dry matter and crude protein digestibilities were not lowered by adding washed or autoclaved manure to a basal concentrate ration (Anthony, 1970). Steers fed basal concentrate or basal concentrate with autoclaved manure gained equally well, whereas steers fed the basal with washed manure gained less, but similar to steers fed a ration of 1 part corn silage to 1 part ear corn. In a second experiment (Anthony, 1970), steers fed the basal gained 1.16 kg/day, while steers fed manure either 40 parts untreated or 40 parts autoclaved manure to 60 parts basal gained 1.00 and 0.99 kg/day. Anthony (1970) concluded that washing or autoclaving manure before mixing it with concentrate did not improve its nutritive value over untreated fresh manure.

Anthony (1969, 1971) recycled feedlot manure by mixing manure with Bermuda hay and ensiled the mixture. The silage, called "wastelage" by Anthony, was 57 parts manure and 43 parts hay, wet weight. On a dry basis, wastelage is composed of about 20 percent manure and 80 percent hay. Some groups of steers fed wastelage have gained faster (1.17 kg/day) than control concentrate-fed groups (1.10 kg/day) (Anthony, 1969). In other trials, ADG has equaled control-fed steers (Anthony, 1969, 1971). However, a general observation was higher feed : gain ratios for steers fed manure-containing rations (Anthony, 1969, 1970, 1971).

Digestibilities of feedlot wastes were determined with sheep (Johnson, 1972). The wastes were characterized by high ash content varying from 35 to 44 percent, most of which was probably of soil origin. The organic matter portion of the waste contained between 20–30 percent crude protein, of which 60–71 percent was digestible. McClure *et al.* (1971) reported a much lower value (less than 40.3 percent) for the digestibility of crude protein in corn-fed cattle feces.

Anthony (1969, 1971) presented data on amino acid profiles of beef cattle manure, and Jones *et al.* (1972) showed similar data for a high-protein material fractionated from manure. The material fractionated by Jones *et al.* (1972) had a crude protein content of about 35 percent. They showed that amino acid composition was superior to cereal grains and comparable to soybean. This same type of high-protein material, however, failed to support rat growth when incorporated in otherwise balanced diets containing 10 and 20 percent crude protein (L. W. Smith and H. H. Sloneker, unpublished data). It is apparent that amino acid profiles are not necessarily indicative of nutritive value.

Poultry Litter

Chance (1965) and Anthony (1967, 1971) reviewed the use of poultry litter in livestock diets. Reports included in these reviews and a report since these reviews (Fontenot *et al.*, 1971) indicate that poultry litter is a good source of nitrogen and minerals for ruminants. Fontenot *et al.* (1971) studied the effect of various processing methods on the chemical composition and sterility of broiler litter. They concluded that treatment with dry heat at 150 °C for 3 hr or longer was the only method that resulted in completely sterilized litter. The large loss of nitrogen that resulted was substantially reduced by acidifying the litter with sulfuric acid prior to heating and with no alteration on nitrogen utilization by lambs.

Sterilized dry poultry manure containing about 30 percent crude protein was fed in conventional diets at levels of 10, 20, and 30 percent for growing pigs. While all the pigs remained healthy, for each 10 percent addition of litter, growth was reduced by 0.02 kg/day and feed efficiency by 0.25 units (Perez-Aleman *et al.*, 1971).

Quisenberry and Bradley (1968) substituted ground litter from replacement pullets and broilers at either 10 or 20 percent of the diet at the expense of soybean meal in a ration for laying hens. Their results indicate that body weight, hen-day egg production, feed efficiency, and egg size were not affected by these isonitrogenous and isocaloric substitutions.

Swine Manure

Diggs *et al.* (1965) reported that finishing swine fed diets containing 15 percent dried swine manure had similar rates of gain and efficiency of feed utilization as those fed a basal control diet. Average daily gains and feed efficiency were both severely depressed by replacing all the soybean meal in corn–soybean oil meal diets with dried swine feces (Orr, 1971). Harmon *et al.* (1971b) demonstrated that weight gains and feed efficiencies of rats fed 4.0 percent freeze-dried oxidation ditch mixed liquor (FDODML) were similar to control-fed rats. Weight gains and feed efficiencies were lower when rats were fed 8.0 or 12.0 percent FDODML of the diet. Harmon *et al.* (1972b) concluded that 25–33 percent of the protein from casein or soybean meal could be replaced with oxidation ditch residue (ODR) protein without reducing rat growth. However, feed : gain ratio increased with increased amounts of ODR in the diets. Although feed intakes were not reduced by feeding ODR, chemical analysis of the feces and nitrogen balance data suggest low digestibilities of energy and protein in ODR by rats (Harmon *et al.*, 1972b). Feeding two parts oxidation ditch mixed liquor per one part dry diet (Harmon *et al.*, 1971a, 1972a) of 12 percent in crude protein from corn–soybean meal resulted in higher ADG (0.475–0.530 kg) and improved gain : feed ratios (0.215 : 0.241) for finishing pigs compared with the same dry ration with water added.

Dairy Cattle Manure

Dairy wastes containing manure plus peanut-hull bedding were fed to 15 Holstein heifers (Smith and Gordon, 1971). Three rations were formulated from corn meal, dehydrated dairy cattle manure (13 percent crude protein), and supplemental urea so that all were isonitrogenous at 15.5 percent crude protein. Average dry matter intakes, growth rates, and feed : gain ratios were not different for the three rations containing from 33 to 67 percent dairy wastes. However, the highest gain (0.47 kg/day) was obtained at the lowest manure content, while the ration highest in manure resulted in a relative saving of 16 percent corn/kg of gain.

Nitrogen in feces from all forage-fed animals was 0–18 percent digestible as determined with sheep (Smith *et al.*, 1969). Dehydrated and pelleted rations entirely of dairy cattle manure were consumed well by mature sheep (4.7 percent of body weight) and maintained positive nitrogen balances (Smith *et al.*, 1971). Thirty-two (Smith *et al.*, 1971) to 48 percent (Thomas *et al.*, 1970) of the nitrogen in dairy cattle manure was digested by sheep.

OTHER ASPECTS THAT MAY INFLUENCE UTILIZATION OF ANIMAL WASTE AS A PROTEIN SOURCE

Digestibilities of dry matter, organic matter, and crude protein of various animal wastes when fed to sheep are presented in Table 5. The relatively low dry matter and organic matter digestibilities may prohibit or limit the use of beef and dairy cattle wastes in rations re-

TABLE 5 *In vivo* Digestibilities of Animal Waste by Sheep

Animals			Digestibility (%)			
From	To	Level Fed	Dry Matter	Organic Matter	N	References
Caged broilers	Sheep	100	54	61	67	L. W. Smith and C. C. Calvert (unpublished)
Caged layers	Sheep	100	57	67	77	Lowman and Knight (1970)
Caged layers	Sheep	32	39	–	58	Thomas *et al.* (1970)
Caged layers	Sheep	20–80	53	64	–	Tinnimet *et al.* (1972)
Poultry litter	Sheep	36	–	–	58	Bhattacharya and Fontenot (1965)
Beef cattle	Sheep	25–40	40	49	67	Johnson (1972)
Beef cattle	Sheep	25–40	35	42	60	Johnson (1972)
Beef cattle	Sheep	25–40	50	56	71	Johnson (1972)
Dairy cattle	Sheep	100	27	–	31	Smith *et al.* (1971)
Dairy cattle	Sheep	39	29	–	48	Thomas *et al.* (1970)
Dairy heifers	Sheep	25–50	22	–	0	Smith *et al.* (1969)

quiring high concentration of digestible energy. Digestibilities of swine waste are expected to be similar to those for beef cattle. Dairy cattle manure is lowest in digestible dry matter and digestible protein, yet it has been shown (Smith *et al.*, 1971) that adult sheep will maintain positive nitrogen balances when fed rations of solely dehydrated and pelleted dairy waste. McInnes *et al.* (1968) report that weaner sheep could be maintained through a drought period with a poultry litter–wheat mixture. Perhaps wastes other than poultry droppings will find greater utility in rations for ruminants maintained at relatively low levels of productivity such as in wintering sheep, beef cattle, and other livestock maintained for reproduction. However, endogenous and exogenous levels of estrogenic materials found in animal wastes (Hurst *et al.*, 1957; Mathur and Common, 1969; T. R. Wrenn, C. C. Calvert, and L. W. Smith, unpublished data) may deter recycling wastes to reproductive livestock.

Ousterhout and Presser (1971) suggest that recycling wet hen manure blended with mash will reduce the manure disposal problem by no more than 25 percent and that there is no noticeable further reduction upon repeated recycling. This observation indicates about a 25 percent digestibility of hen-manure dry matter and is in close agreement with observations of Nesheim (1972). The metabolizable energy (ME) value for poultry waste has been reported to be 200–480 kcal/kg (Neisheim, 1972), 1,100 kcal/kg (Pryor and Connor, 1964), and 1,320 kcal/kg (Polin *et al.*, 1971). Hodgetts (1971) reported two energy values for hen manure, 1,912 kcal/kg using a "classical" determination and 873 kcal/kg using an "available carbohydrate" procedure described by Bolton (1967). The correct ME value of hen manure for poultry is still uncertain, although Pryor and Connor (1964) showed that poultry feces had a ME value of about 30 percent of the grains from which it originated and suggested that an even lower ME value would have been obtained had the original diet been balanced. Thus, the data of Nesheim (1972), Ousterhout and Presser (1971), and Pryor and Connor (1964) offer strong evidence that the more realistic ME value of poultry waste falls between 480 and 873 kcal/kg. Hen manure does not seem to be a desirable ingredient for high-level feeding in poultry diets because of the probably low ME and proportionately low amino acid nitrogen content of hen manure in spite of the apparent favorable performance of hens fed this ingredient at low levels. Ruminants (Lowman and Knight, 1970) have been shown to digest dried-poultry-manure dry matter and crude protein and utilize poultry-manure nitrogen similar to when these constituents were from conventional sources of nitrogen (El-Sabban *et al.*, 1970).

The apparent digestibility of crude protein by ruminants increases as the concentration of protein increases in rations of forages, concentrates, or mixtures (Holter and Reid, 1959; Blaxter and Mitchell, 1948; Dijkstra, 1966; Anderson and Lamb, 1967). The relationship is highly correlated ($r = 0.96$ to 0.99) and is a method to predict digestible protein from crude protein content. The slope of the regression of digestible protein on crude protein estimates true protein digestibility. Protein in most feeds is in the range of 85–90 percent true digestibility. Figure 1 contains this relationship for rations containing 20–100 percent poultry manure and poultry litter (Lowman and Knight, 1970; El-Sabban *et al.*, 1970; McInnes *et al.*, 1968; Bhattacharya and Fontenot, 1965, 1966; Tinnimit *et al.*, 1972; Bull and Reid, 1971). The open line is for forages (Holter and Reid, 1959). Similarly, rations of 13–100 percent cattle manure (Tinnimit *et al.*, 1972; Anthony, 1970; Johnson, 1972; Smith *et al.*, 1969, 1971) are in Figure 2. These few data estimate the true protein digestibility at 87 percent for poultry waste and 94 percent for cattle waste rations. Thus, supplementation of conventional feeds with animal waste nitrogen does not appear to depress total ration protein digestibility. Since no data points diverged greatly from the regression line, it is suggested that these values are representative of true digestibility of nitrogen in animal waste.

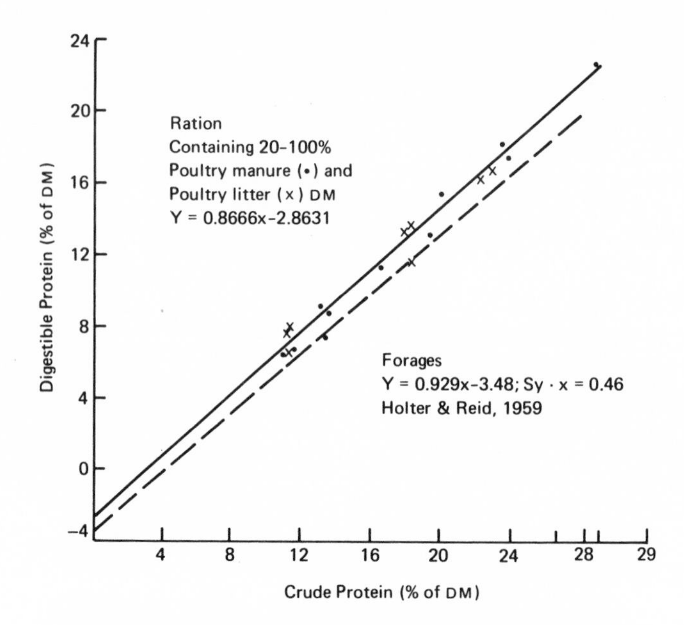

FIGURE 1 The relationship between the crude and digestible protein concentrations of rations composed of 20-100 percent poultry manure and poultry litter dry matter compared with forage.

Methods of handling and processing animal waste for feeding can result in adverse effects on its chemical composition and thus possibly on nutritive value. Fontenot *et al.* (1971) have shown nitrogen losses due to drying litter at 150 °C that were partially prevented by pretreating with acid. Sheppard *et al.* (1971) demonstrated that nitrogen losses from hen manure were directly related to temperatures of drying. Nitrogen and energy losses appear to be related to both temperature and time required to dry poultry manure (Shannon and Brown, 1969). Losses of nitrogen and energy during drying have been observed for pig feces (Saben and Bowland, 1971), and such losses must be expected for other kinds of animal waste. Data are not available to suggest an adverse effect of drying temperature on nutritive values of waste protein. However, with sufficient available carbohydrate, conditions would appear favorable for heat damage of protein (Anantharaman and Carpenter, 1971; Bjarnason and Carpenter, 1970).

Fontenot *et al.* (1971), Chance (1965), and El-Sabban *et al.* (1969) have pointed out the wide variation in chemical composition of poultry litters. Other wastes, even though not influenced by as many variables as litter, should also be expected to vary in composition. The

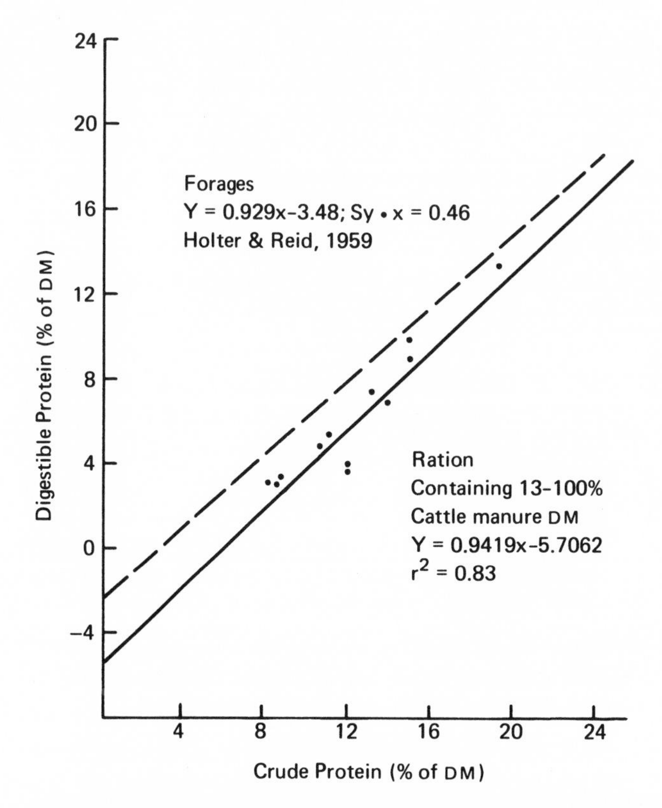

FIGURE 2 The relationship between the crude and digestible protein concentrations of rations composed of 13–100 percent cattle manure dry matter compared with forage.

composition of feces from all livestock are largely affected by the composition of the diet. Many additional factors may affect composition and nutritive value while processing for feed. Thus, an obvious need is for specific identification and descriptions of wastes for recycling.

Numerous experiments have been conducted to study the feeding value of various animal waste for different livestock species. Some of these feeding trials also included concurrent taste panel studies of consumer products derived from animals fed the wastes. Eggs (Flegal and Zindel, 1971), beef (El-Sabban *et al.*, 1970; Fontenot *et al.*, 1971; Rhodes, 1971), and milk (Bucholtz *et al.*, 1971) from animals fed dried poultry waste or litter could not be distinguished from control products.

From the data of Fontenot *et al.* (1971) and El-Sabban *et al.*

(1970), it may be concluded that accumulations of chlorinated hydrocarbon pesticides did not appear to be of any consequence or a problem peculiar to poultry litter or droppings. Similar information is needed for other wastes.

Broiler litter was rendered free of pathogenic organisms by various heat treatments (Fontenot *et al.*, 1971). Alexander *et al.* (1968) provided evidence on the presence of pathogenic organisms in poultry litter fed to livestock in Canada and suggests that 1–2 months storage may be sufficient to destroy *Salmonella* present in litter. Whether or not storage would be adequate to destroy other pathogenic bacteria is not known (Alexander *et al.*, 1968). Messer *et al.* (1971) also demonstrated that *Salmonella* and *Arizona* sp. are not highly resistant to heat in poultry litter of normal moisture content and concluded a heat process may be feasible for their elimination.

More information is needed on the effect of recycling medicants in some wastes on animals and animal products. Various controls can be instituted. One approach would be not to use medicants in recycling systems. If a medicant must be used, perhaps recycling should be avoided. Also, wastes could be withheld from diets so as to allow adequate tissue clearance.

Studies are needed to determine whether or not closed, continuous recycle systems are feasible. Present information suggests that possible accumulations of minerals, peptidoglycan, and lignocellulose would occur without preventive measures. Peptidoglycan and lignocellulose are nonnutritive residues for monogastrics, while only lignocellulose is nonnutritive for ruminants. Intermediate microbial, chemical, or physical processing recovery methods could prevent these accumulations in closed recycling systems. Perhaps a more logical approach would be to develop information for feasible ratios of recycled to conventional nutrients for continuous systems similar to that purposed by Anthony (1971).

SUMMARY

Animal waste nitrogen is utilized when fed in diets of livestock. Ruminants appear to be better qualified to utilize animal waste nitrogen than other species. Droppings from caged poultry, especially layers, appear to be the most suitable for recycling to ruminants, based upon nutritive value and freedom from medicants that often occur in other wastes. Continually evolving technological advances will probably result in physical and fermentative advances for conversion of animal waste nitrogen into products of even higher nutritive value for livestock feeding.

REFERENCES

Alexander, D. C., J. A. J. Carriere, and K. A. McKay. 1968. Bacteriological studies of poultry litter fed to livestock. Can. Vet. J. 9:127–131.

Allison, M. J. 1969. Biosynthesis of amino acids by ruminal microorganisms. J. Anim. Sci. 29:797–807.

Anantharaman, K., and K. J. Carpenter. 1971. Effects of heat processing on nutritional value of groundnut products. J. Sci. Food Agric. 22:412–417.

Anderson, M. J., and R. C. Lamb. 1967. Predicting digestible protein from crude protein. Proc. West. Sect. Am. Soc. Anim. Sci. 18:81–86.

Anthony, W. B. 1966. Utilization of animal waste as feed for ruminants, p. 109–112. *In* Management of farm animal wastes. Proceedings National Symposium, Publ. SP-0366, Am. Soc. Agric. Eng., St. Joseph, Mich.

Anthony, W. B. 1967. Review of studies of feeding poultry litter to livestock. Proceedings Poultry Litter Seminar, Auburn University, Auburn, Ala. 136 p.

Anthony, W. B. 1969. Cattle manure: Reuse through wastelage feeding, p. 105–113. *In* Animal waste management. Proceedings Conference on Animal Waste Management, Cornell University, Ithaca, N.Y.

Anthony, W. B. 1970. Feeding value of cattle manure for cattle. J. Anim. Sci. 30:274–277.

Anthony, W. B. 1971. Cattle manure as feed for cattle, p. 293–296. *In* Livestock waste management and pollution abatement. Proceedings International Symposium on Livestock Wastes, Columbus, Ohio, Am. Soc. Agric. Eng., St. Joseph, Mich.

Anthony, W. B., and R. Nix. 1962. Feeding potential of reclaimed fecal residue. J. Dairy Sci. 45:1538–1539.

Becker, D. E., A. H. Jensen, and B. G. Harmon. 1963. Balancing swine rations. Univ. Ill. Coll. Agric. Coop. Ext. Serv. Circ. 866. 32 p.

Belasco, I. J. 1954. New nitrogen feed compounds for ruminants—A laboratory evaluation. J. Anim. Sci. 13:601–610.

Benedict, G., and E. G. Ritzman. 1923. Undernutrition in steers—Its relation to metabolism, digestion, and subsequent realimentation. Carnegie Institute, Washington, D.C. Publ. 324. 333 p.

Bergen, W. G., D. B. Purser, and J. H. Cline. 1968. Determination of limiting amino acids of rumen-isolated microbial proteins fed to rats. J. Dairy Sci. 51:1698–1699.

Bhattacharya, A. N., and J. P. Fontenot. 1965. Utilization of different levels of poultry litter nitrogen by sheep. J. Anim. Sci. 24:1174–1178.

Bhattacharya, A. N., and J. P. Fontenot. 1966. Protein and energy value of peanut hull and wood shaving poultry litters. J. Anim. Sci. 25:367–371.

Bjarnason, J., and K. J. Carpenter. 1970. Mechanisms of heat damage in proteins. 2. Chemical changes in pure proteins. Br. J. Nutr. 24:313–329.

Blaxter, K. L., and H. H. Mitchell. 1948. The factorization of the protein requirement of ruminants and of the protein values of feeds, with particular reference to the significance of the metabolic fecal nitrogen. J. Anim. Sci. 7:351–372.

Bolton, W. 1954. Poultry, p. 110–112, *In* John Hammond [ed.] Progress in the physiology of farm animals, Vol. 1. Butterworths Scientific Publications, London.

Bolton, W. 1967. Poultry nutrition. Bull. 174, Her Majesty's Stationery Office, London.

Breon, W. S. 1939. The determination of protein digestibility in the fowl. M.S. thesis. Cited in [J. W. White, F. J. Holben, and A. C. Richer] 1944. Production, composition, and value of poultry manure. Pa. St. Coll. Sch. Agric. Bull. 469. p. 21.

Bryant, M. P., and I. M. Robinson. 1962. Some nutritional characteristics of predominate culturable ruminal bacteria. J. Bacteriol. 84:605–614.

Bryant, M. P., and I. M. Robinson. 1963. Apparent incorporation of ammonia and amino acid carbon during growth of selected species of ruminal bacteria. J. Dairy Sci. 46:150–154.

Bucholtz, H. F., H. E. Henderson, J. W. Thomas, and H. C. Zindel. 1971. Dried animal waste as a protein supplement for ruminants, p. 308–310. *In* Livestock waste management and pollution abatement. Proceedings International Symposium on Livestock Wastes, Columbus, Ohio, Am. Soc. Agric. Eng., St. Joseph, Mich.

Bull, L. S., and J. T. Reid. 1971. Nutritive value of chicken manure for cattle, p. 297–300. *In* Livestock waste management and pollution abatement. Proceedings International Symposium on Livestock Wastes, Columbus, Ohio, Am. Soc. Agric. Eng., St. Joseph, Mich.

Calvert, C. C., N. O. Morgan, and H. J. Eby. 1971. Biodegraded hen manure and adult house flies: Their nutritional value to the growing chick, p. 319–320. *In* Livestock waste management and pollution abatement. Proceedings International Symposium on Livestock Wastes, Columbus, Ohio, Am. Soc. Agric. Eng., St. Joseph, Mich.

Chance, C. M. 1965. Non-protein nitrogen and poultry litter in ruminant diets. Proceedings Maryland Nutrition Conference, Univ. Md., College Park. 99 p.

Committee on Animal Nutrition, National Research Council. 1968. Nutrient requirements of sheep. Fourth rev. ed. National Academy of Sciences, Washington, D.C.

Conrad, H. R., J. W. Hibbs, and A. D. Pratt. 1960. Nitrogen metabolism in dairy cattle, Ohio Agric. Exp. Stn. Res. Bull. 861. 48 p.

Cuthbertson, Sir David. 1969. Animal product and world needs, p. 1311–1329. *In* Sir David Cuthbertson [ed.], Nutrition of animals of agricultural importance, Vol. 17, Part 2. Pergamon Press, N.Y.

Diggs, B. G., B. Baker, Jr., and F. G. James. 1965. Value of pig feces in swine finishing rations. J. Anim. Sci. 24:291.

Dijkstra, N. D. 1966. Estimation of the nutritive value of fresh roughage. Proceedings X International Grassl. Congress, Univ. Helsinki, Finland. 1015 p.

El-Sabban, F. F., T. A. Long, R. F. Gentry, and D. E. H. Frear. 1969. The influence of various factors on poultry litter composition, p. 340–346. *In* Animal waste management. Proceedings Conference on Animal Waste Management, Cornell University, Ithaca, N.Y.

El-Sabban, F. F., J. W. Bratzler, T. A. Long, D. E. H. Frear, and R. F. Gentry. 1970. Value of processed poultry waste as a feed for ruminants. J. Anim. Sci. 31:107–111.

Ellis, W. C., and K. L. Bleichner. 1969. Synthesis and digestion of nucleic acids in the gastrointestinal tract of sheep. Fed. Proc. 28(2):623.

Ellis, W. C., and W. H. Pfander. 1965. Rumen microbial polynucleotide synthesis and its possible role in ruminant nitrogen utilization. Nature 205: 974–975.

Flegal, C. J., and D. A. Dorn. 1971. The effects of continually recycling dehydrated poultry wastes (DPW) on the performance of SCWL laying hens—A preliminary report. *In* Poultry pollution: Problems and solutions. Mich. St. Univ. Agric. Exp. Stn. Res. Rep. 152:45–48.

Flegal, C. J., and H. C. Zindel. 1970. The result of feeding dried poultry waste to laying hens on egg production and feed conversion. *In* Poultry pollution: Problems and solutions. Mich. St. Univ. Res. Rep. 117:29–33.

Flegal, C. J., and H. C. Zindel. 1971. Dehydrated poultry waste (DPW) as a feedstuff in poultry rations, p. 305–307. *In* Livestock waste management and pollution abatement. Proceedings International Symposium on Livestock Wastes, Columbus, Ohio, Am. Soc. Agric. Eng., St. Joseph, Mich.

Fontenot, J. P., K. E. Webb, Jr., B. W. Harmon, R. E. Tucker, and W. E. C. Moore. 1971. Studies of processing, nutritional value, and palatability of broiler litter for ruminants, p. 301–304. *In* Livestock waste management and pollution abatement. Proceedings International Symposium on Livestock Wastes, Columbus, Ohio. Am. Soc. Agric. Eng., St. Joseph, Mich.

Harmon, B. G., D. L. Day, A. H. Jensen, and D. H. Baker. 1971a. Liquid feeding of oxidation ditch mixed liquor to swine. J. Anim. Sci. 33:1149.

Harmon, B. G., D. L. Day, A. J. Jensen, and D. H. Baker. 1971b. Nutritive value of oxidation ditch mixed liquor for rats. J. Anim. Sci. 33:1149.

Harmon, B. G., D. L. Day, D. H. Baker, S. E. Curtis, and A. H. Jensen. 1972a. Harvesting nutrients from swine waste. Ill. Pork Ind. Field Day Rep. AS-661-g. 23 p.

Harmon, B. G., D. L. Day, A. H. Jensen, and D. H. Baker. 1972b. Nutritive value of aerobically sustained swine excrement. J. Anim. Sci. 34:403–407.

Hart, E. B., E. V. McCollum, H. Steenbock, and G. C. Humphrey. 1911. Physiological effect on growth and reproduction of rations balanced from restricted sources. Univ. Wisc. Exp. Stn. Res. Bull. 17:180–183.

Hatfield, E. E., R. M. Forbes, A. L. Neumann, and U. S. Garrigus. 1955. A nitrogen balance study with steers using urea, biuret and soybean oil meal as sources of nitrogen. J. Anim. Sci. 14:1206.

Hobson, P. N. 1969. Microbiology of digestion in ruminants and its nutritional significance, p. 65. *In* Sir David Cuthbertson [ed.], Nutrition of animals of agricultural importance. Part 1. The science of nutrition of farm livestock. Pergamon Press, New York.

Hodgetts, B. 1971. The effects of including dried poultry waste in the feed of laying hens, p. 311–313. *In* Livestock waste management and pollution abatement. Proceedings International Symposium on Livestock Wastes, Columbus, Ohio, Am. Soc. Agric. Eng., St. Joseph, Mich.

Holter, J. A., and J. T. Reid. 1959. Relationship between the concentrations of crude protein and apparently digestible protein in forages. J. Anim. Sci. 18:1339–1349.

Hoogenraad, N. J., and F. J. R. Hird. 1970. The chemical composition of rumen bacteria and cell walls from rumen bacteria. Br. J. Nutr. 24:119–127.

Hoogenraad, N. J., F. J. R. Hird, R. G. White, and R. A. Leng. 1970. Utilization of ^{14}C-labelled *Bacillus subtilis* and *Escherichia coli* by sheep. Br. J. Nutr. 24:129–144.

Hurst, R. O., A. Kukis, and J. F. Bendell. 1957. The separation of oestrogens from avian droppings. Can. J. Biochem. Physiol. 35:637–640.

Johnson, B. C., T. S. Hamilton, and W. B. Robinson. 1944. On the mechanism of non-protein nitrogen utilization by ruminants. J. Anim. Sci. 3:287–298.

Johnson, R. R. 1972. Digestibility of feedlot waste. Okla. Agric. Exp. Stn. Misc. Publ. 87:62–65.

Jones, R. W., J. H. Sloneker, and G. E. Inglett. 1972. Recovery of animal feed from cattle manure. Proceedings 18th Annual Technical Meeting. Institute Environmental Sciences, N.Y.

Jurtshuk, P., Jr., R. N. Doetsch, and J. C. Shaw. 1955. Anaerobic purine dissimilation by washed suspension of bovine rumen bacteria. J. Dairy Sci. 41:190–202.

Keys, J. E., Jr., P. J. Van Soest, and E. P. Young. 1969. Comparative study of the digestibility of forage cellulose and hemicellulose in ruminants and non-ruminants. J. Anim. Sci. 29:11–15.

Looper, C. G., and O. T. Stallcup. 1958. Release of ammonia nitrogen from uric acid, urea, and certain amino acids in the presence of rumen microorganisms. J. Dairy Sci. 41:729.

Loosli, J. K., H. H. Williams, W. E. Thomas, F. H. Ferris, and L. A. Maynard. 1949. Synthesis of amino acids in the rumen. Science 110:144–145.

Lowman, B. G., and D. W. Knight. 1970. A note on the apparent digestibility of energy and protein in dried poultry excreta. Anim. Prod. 12:525–528.

McClure, K. E., R. D. Vance, E. W. Klosterman, and R. L. Preston. 1971. Digestibility of feces from cattle fed finishing rations. J. Anim. Sci. 33:292.

McGilliard, A. D. 1972. Modifying proteins for maximum utilization in the ruminants. J. Am. Oil Chem. Soc. 49:57–62.

McInnes, P., P. J. Austin, and D. L. Jenkins. 1968. The value of a poultry litter and wheat mixture in the drought feeding of weaner sheep. Aust. J. Exp. Agric. Anim. Husb. 8:401–404.

McNaught, M. L., J. A. B. Smith, K. M. Henry, and S. K. Kon. 1950. The utilization of non-protein nitrogen in the bovine rumen. 5. The isolation and nutritive value of a preparation of dried rumen bacteria. J. Biochem. 46:32–36.

McNaught, M. L., E. C. Owen, K. M. Henry, and S. K. Kon. 1954. The utilization of non-protein nitrogen in the bovine rumen. 8. The nutritive value of the protein of preparations of dried rumen bacteria, rumen protozoa, and brewers yeast for rats. J. Biochem. 56:151–156.

Mason, V. C. 1969. Some observations on the distribution and origin of nitrogen in sheep feces. J. Agric. Sci. Camb. 73:99–111.

Mason, V. C., and G. Milne. 1971. The digestion of bacterial mucopeptide constituents in the sheep. 2. The digestion of musmic acid. J. Agric. Sci. Camb. 77:99–101.
Mason, V. C., and R. Palmer. 1971. Studies on the digestibility and utilization of the nitrogen of irradiated rumen bacteria by rats. J. Agric. Sci. Camb. 76:567–572.
Mason, V. C., and F. White. 1971. The digestion of bacterial mucopeptide constituents in the sheep. 1. The metabolism of 2,6-diaminopimelic acid. J. Agric. Sci. Camb. 77:91–98.
Mathur, R. S., and R. H. Common. 1969. A note on the daily urinary excretion of estradiol-17β and estrone by the hen. Poult. Sci. 48:100–104.
Messer, J. W., J. Lovett, G. K. Murthy, A. J. Wehby, M. L. Schafer, and R. B. Read. 1970. An assessment of some public health problems resulting from feeding poultry litter to animals. Microbiological and chemical parameters. Poult. Sci. 50:874–881.
Morris, S., and S. C. Ray. 1939a. CXLVIII. The effect of a phosphorus deficiency on the protein and mineral metabolism of sheep. Biochem. J. 33: 1209–1216.
Morris, S., and S. C. Ray. 1939b. CXLIX. The fasting metabolism of ruminants. Biochem. J. 33:1217–1230.
Morrison, F. B. 1957. Feeds and feeding. 22 ed. Morrison Publ. Co., Ithaca, N.Y. 1165 p.
Nesheim, M. C. 1972. Evaluation of dehydrated poultry manure as a potential poultry feed ingredient, p. 301–308. *In* Waste management research. Cornell Agricultural Waste Management Conference, Syracuse, N.Y. Graphics Mgmt. Corp., Washington, D.C.
O'Dell, G. L., W. D. Woods, O. A. Laerdal, A. M. Jeffay, and J. E. Savage. 1960. Distribution of the major nitrogenous compounds and amino acids in the chicken urine. Poult. Sci. 39:426–432.
Oltjen, R. R., L. L. Slyter, A. S. Kozak, and E. E. Williams, Jr. 1968. Evaluation of urea, biuret, urea phosphate, and uric acid as NPN sources for cattle. J. Nutr. 94:193–202.
Oltjen, R. R., D. A. Dinius, M. I. Poos, and E. E. Williams, Jr. 1972. Na urate, 25% urate and uric acid as NPN sources for beef cattle. J. Anim. Sci. 35:272.
Orr, D. E. 1971. Recycling dried waste to finishing pigs. Mich. St. Univ. Agric. Exp. Stn. Rep. Swine Res. 148:63–68.
Ousterhout, L. E., and R. H. Presser. 1971. Increased feces production from hens being fed poultry manure. Poult. Sci. 50:1614.
Perez-Aleman, S., D. C. Dempster, P. R. English, and J. H. Topps. 1971. A note on dried poultry manure in the diet of the growing pig. Anim. Prod. 13:361–364.
Pilgrim, A. F., F. V. Gray, R. A. Weller, and C. B. Belling. 1970. Synthesis of microbial protein from ammonia in the sheep's rumen and the proportion of dietary nitrogen converted into microbial nitrogen. Br. J. Nutr. 24:589–598.
Pisone, U., and J. J. Begin. 1971. Recycling animal waste through poultry. II. Dried poultry manure. Ky. Anim. Sci. Res. Prog. Rep. 196:34–35.

Polin, D., S. Varghese, M. Neff, M. Gomez, C. J. Flegal, and H. C. Zindel. 1971. The metabolizable energy value of dried poultry waste. Poultry pollution: Research results. Mich. St. Univ. Agric. Exp. Stn. Rep. 152:32–44.

Pryor, W. J., and J. K. Connor. 1964. A note on the utilization by chickens of energy from faeces. Poult. Sci. 43:833–834.

Quisenberry, J. H., and J. W. Bradley. 1968. Nutrient recycling. Second National Poultry Litter and Waste Management Seminar. College Station, Texas. p. 96–106.

Reed, F. M., R. J. Moir, and E. J. Underwood. 1949. Ruminal flora studies in the sheep. 1. The nutritive value of rumen bacterial protein. Aust. J. Sci. Res. 2:304–317.

Reid, J. T. 1970. The future role of ruminants in animal production, p. 1–22. *In* A. T. Phillipson [ed.], Physiology of digestion and metabolism in the ruminant. Oriel Press, Ltd., New Castle Upon Tyne, England.

Rhodes, D. N. 1971. Flavour of beef fed on dried poultry waste. J. Sci. Food Agric. 22:336.

Rusnak, J. J., T. A. Long, and T. B. King. 1966. Value of hydrolyzed poultry waste as a protein supplement for beef cattle. Penn. St. Univ. Anim. Sci. Mimeo. 11 p.

Saban, H. S., and J. P. Bowland. 1971. Comparative evaluation of some techniques used in determinations of nitrogen and energy content of feces from pigs. Can. J. Anim. Sci. 51:793–799.

Salton, M. R. J., and J. G. Pavlik. 1960. Studies of the bacterial cell wall. VI. Wall composition and sensitivity to lysozyme. Biochem. Biophys. Acta 39:398–407.

Scott, M. L., M. C. Nesheim, and R. J. Young. 1969. Nutrition of the chicken. M. L. Scott & Associates, Ithaca, N.Y. 511 p.

Shannon, D. W. F., and W. O. Brown. 1969. Losses of energy and nitrogen on drying poultry excreta. Poult. Sci. 48:41–43.

Sheppard, C. C., C. J. Flegal, D. Dorn, and J. L. Dale. 1971. The relationship of drying temperature to total crude protein in dried poultry waste. Poultry pollution: Research results. Mich. St. Univ. Agric. Exp. Stn. Res. Rep. 153: 12–16.

Smith, L. W. 1965. Observations on the flora of the alimentary tract of animals and factors affecting its composition. J. Pathol. Bacteriol. 89:95–122.

Smith, L. W., and C. H. Gordon. 1971. Dairy cattle manure—Corn meal rations for growing heifers. J. Anim. Sci. 33:300.

Smith, R. H., and A. B. McAllan. 1970. Nucleic acid metabolism in the ruminant. 2. Formation of microbial nucleic acids in the rumen in relation to the digestion of food nitrogen and the fate of dietary nucleic acids. Br. J. Nutr. 24:545–556.

Smith, L. W., H. K. Goering, and C. H. Gordon. 1969. Influence of chemical treatments upon digestibility of ruminant feces, p. 88–97. *In* Animal waste management. Conference on Animal Waste, Cornell University, Ithaca, N.Y.

Smith, L. W., H. K. Goering, and C. H. Gordon. 1971. Nutritive evaluations of untreated and chemically treated dairy cattle wastes, p. 314–318. *In* Livestock waste management and pollution abatement. Proceedings International Symposium on Livestock Wastes, Columbus, Ohio, Am. Soc. Agric. Eng., St. Joseph, Mich.

Stangel, H. J., R. R. Johnson, and A. Spellman. 1963. Urea and non-protein nitrogen in ruminant nutrition. [ed.] H. J. Stangel. Allied Chemical Corp., 2nd Ed. 489 p.

Stekol, J. A. 1936. Comparative studies in the sulfur metabolism of the dog and pig. J. Biol. Chem. 113:675–682.

Taiganides, E. P., and R. L. Stroshine. 1971. Impact of farm animal production and processing on the total environment, 95–98. *In* Livestock waste management and pollution abatement. Proceedings International Symposium on Livestock Wastes, Columbus, Ohio. Am. Soc. Agric. Eng., St. Joseph, Mich.

Thomas, J. W., Yu Yu, and J. A. Hoffer. 1970. Digestibility of paper and dehydrated feces. J. Anim. Sci. 31:255.

Thomas, J. W., Yu Yu, P. Tinnimit, and H. C. Zindel. 1972. Dehydrated poultry waste as a feed for milking cows and growing sheep. J. Dairy Sci. 55:1261–1265.

Tinnimit, P., Yu Yu, K. McGuffey, and J. W. Thomas. 1972. Dried animal waste as a protein supplement for sheep. J. Anim. Sci. 35:431–435.

Train, R. E., R. Cahn, and G. J. McDonald. 1970. Environmental quality. The first annual report of the council on environmental quality, Washington, D.C. GPO, Washington, D.C. 326 p.

Virtanen, A. I. 1966. Milk production of cows on protein-free feed. Science 153:1603–1614.

Waldo, D. R. 1968. Symposium: Nitrogen utilization by the ruminant. Nitrogen metabolism in the ruminant. J. Dairy Sci. 51:265–275.

Wehunt, K. E., H. L. Fuller, and H. M. Edwards, Jr. 1960. The nutritional value of hydrolyzed poultry manure for broiler chickens. Poult. Sci. 39:1057–1063.

Willingale, J. M., and C. A. E. Briggs. 1955. The normal intestinal flora of the pig. II. Quantitative bacteriological studies. J. Appl. Bacteriol. 18:284–293.

Zebrowska, T. 1968. The course of digestion of different food proteins in the rat. Fractionation of the nitrogen in intestinal contents. Br. J. Nutr. 22:483–491.

IV
Summary

W. M. Beeson

SUMMARY

Protein is one of the major limiting nutrients for the balancing of rations for the production of livestock and poultry for human consumption. These papers have reviewed the potential sources of protein for animal production and have considered how we may increase the quantity and improve the biological value of protein for animal production.

CEREAL-GRAIN PROTEIN

Historically, cereal grains have been developed for high yields and primarily as a source of energy rather than for quantity and quality of protein. In the past decade, however, considerable research emphasis has been placed on improving the amino acid pattern and quantity of protein in cereal grains. All of the cereal grains, except oats, show a negative relationship between the quantity of protein in the grain and the biological quality of the protein. In barley and corn, mutant endosperm types have been found that markedly increase the level of lysine and thus improve the biological value of protein for monogastric animals. There is a good possibility that similar mutant genes will be found for the improvement of the protein quality of sorghum and wheat. New species, such as *Triticale,* have a higher level of lysine than wheat and thus provide a better quality of protein. It appears that cereal grains will make more contributions to the supply of essential amino acids for animal and man.

SEPARATION OF PROTEIN FROM FIBER IN FORAGE CROPS

Green leaf crops, especially alfalfa, are the major source of protein for livestock in this country. The total hay crop is about 128 million tons, of which 75 million tons is alfalfa hay. This alfalfa hay crop contains as much protein (11.2 million tons) as the entire 1.14 billion bushel soybean crop or the 4.1 billion bushel corn crop. On a per acre basis, alfalfa produced from 1,500–4,000 lb of crude protein, and soybeans averaged 1,433 lb of protein.

By air separation, dehydrated alfalfa has been separated into 44.3 percent of a leaf-rich fraction containing 23.0 percent protein and 24.8 percent fiber and 55.7 percent of a stem-rich fraction with 11.5 percent protein and 41.0 percent fiber. In feeding tests with beef cattle and sheep, the alfalfa stem fraction was equal in value to alfalfa hay.

A dual-cutting system, which consisted of cutting and collecting the top part of the plant to segregate a protein-rich fraction and harvesting the bottoms of the plants separately to obtain the low-protein–high-fiber fraction, was compared with the air-separation technique. In the dual-cutting system, 36.0 percent of a supergrade product was produced containing about 30.0 percent crude protein and 15.0 percent fiber. The stem fraction from the bottom half of the plant ran about 15.0 percent protein and 35.0 percent fiber. The practicality of the dual-cutting system has not yet been proven.

A method is described for fractionating alfalfa by a juicing and dehydration process. The specific research objectives were reduction of costs of dehydrated alfalfa by mechanical dewatering, recovery of a high-protein–high-xanthophyll product (PRO-XAN) designed primarily as a pigmentation supplement for poultry, and recovery of solubles fraction from the coagulation step as a molasses or unidentified growth-factor concentrate.

MEAT AND BLOOD BY-PRODUCTS

By-products from the processing of livestock and poultry for human consumption offer a real potential for increasing our supplies of proteins for animals, and, in some cases, for humans. At present, much of the animal by-product protein material is either discarded as a waste material or is improperly processed, resulting in a protein of poor biological value. Blood proteins can be produced in a highly purified form with excellent nutritional value. Protein fractions that have been isolated from blood are excellent sources of lysine and tryptophan. With appropriate processing and blending, animal offal proteins can be made into protein supplements of high nutritive quality for balancing livestock and poultry rations. Collagenous proteins, which are deficient in essential amino acids, can be converted to single-cell protein (SCP) of high nutritive quality by fermentation. Keratin proteins from hair and feathers can be hydrolyzed to make their constituent amino acids almost completely available to poultry.

ADVANCES IN OILSEED PROTEIN UTILIZATION

Oilseed protein sources constitute three fourths of the available protein supplies at present. Protein supplies from oilseeds may be increased by two methods: (1) increasing production of existing oilseed protein sources and (2) changing utilization patterns that will encourage the distribution of protein supplies to areas of greatest need. Soybean meal makes up approximately 88 percent of the total oilseed meal produced in the United States, followed in order by cottonseed meal, linseed, peanut, and copra. The factors that favor soybean meal as a source of protein are: (1) soybeans are a good cash crop; (2) amino acid composition of soybean meal balances the deficiencies in grains for monogastric animals; and (3) lack of plant toxins in soybean meal has also allowed for unlimited usage in livestock rations. Increased yields of soy protein per acre could do much to satisfy the world's need for a relatively cheap protein source.

Unlike soybean meal, cottonseed meal is affected both by supply and by utilization decisions. Cottonseed meal is a by-product, and the supply is directly influenced by the factors affecting cotton production. Cottonseed meal is used principally in the rations of ruminant animals because of its amino acid deficiencies and the presence of gossypol. Since research has defined the limits of gossypol permissible in rations of nonruminant animals, this has increased the use of cottonseed meal and resulted in its evaluation on a nutritional basis.

Other oilseed meals that have made a minor contribution to our protein supplies are linseed meal, peanut meal, and copra meal.

MINOR PLANTS AS PROTEIN SOURCES

Oilseeds, including the oil-bearing tree fruits, supply more than one half of the world's fat and oils and most of the supplementary protein for livestock. Soybeans, cottonseed, and peanuts are the leading oilseeds and represent more than 80 percent of the world's supply of oilseed protein. Some of the minor sources of protein from oil-bearing seeds are sunflower meal, rapeseed meal, sesame meal, coconut meal, and safflower meal. Sunflower meal is most efficiently used by combining it with either soybean meal or fish meal to compensate for its deficiency in lysine. Rapeseed meal has a good balance of essential amino acids, but the presence of thioglucosides limits the amount that can be incorporated in the diet. Sesame meal is an excellent source of methionine and tryptophan, but needs to be blended with such other sources of protein as soybean meal or fish meal, which are good sources of lysine. Coconut meal is deficient in lysine and also has other characteristics that limit its use as the only source of supplementary protein for swine. The protein of safflower meal is a very poor source of lysine and is somewhat deficient in methionine and cystine. Due to its high fiber content and poor quality protein, safflower meal should not be used at a higher level than 7–8 percent in a diet for growing pigs.

The grain legumes, such as soybeans and peanuts, contribute significantly to the total supply of feed proteins; however, some of the other grain legumes, such as field peas, chickpeas, field beans, beans, cowpeas, and pigeon peas, contribute only in a limited way to the total protein supply.

FISH PRODUCTS FOR ANIMAL FEEDING

The bulk of the domestically produced fish meal comes from menhaden. Other contributing species are tuna, mackerel, herring, and anchovy. Since 1950, the 10-year average supply of fish meal in the United States has risen from 250,000 tons (1950–1959) to about 400,000 tons (1960–1969). Using lean fish (3–4 percent fat), a process has been developed for the production of a fish-protein concentrate for human feeding. Research is under way to modify the process to use fatty species of fish such as menhaden, anchovy, and herring as a source of raw material. If the fish-protein concentrate industry develops, it would reduce the supply of fish meal available for feeding livestock. Fish muscle is being studied as a partial replacer of the lean beef used as emulsifiers and nutritional protein sources in sausage products. Texture, flavor, yield, and resistance to spoilage have all been found to be satisfactory when 15 percent of the lean meat portion of the sausage was replaced by an equal weight of fish flesh. Fish-muscle proteins are presently being modified to be essentially flavorless and odorless, soluble in water, and extremely effective as emulsifying agents. These products have potential use as coffee whiteners, dessert toppings, and a partial replacement for egg whites. Recovery of wastes from fish-processing plants to reduce pollution of the environment could represent a source of low protein (40 percent)–high ash fish meal that could be used in animal feeds.

SINGLE-CELL PROTEIN

Three major sources of industrial wastes that may be adapted to the production of SCP are, in order of magnitude, chemical-related industries, paper and allied products, and food industries. Often, the cost of production of SCP from these waste products is too high to justify the process unless part of the cost is charged against pollution control. The methods of producing SCP vary greatly, ranging from simple incubation to very elaborate systems of production with complete environmental and nutrient control. Some of the problems that must be considered in the evaluation of SCP are palatability, digestibility, nucleic acid content, toxic or harmful residues, and amino acid content. Usually, SCP is low in cystine and methionine. Other amino acids that may be marginal are lysine and isoleucine. Many SCP products are equivalent to casein or soybean meal when supplemented with methionine. In this paper, the importance of SCP production is established, and the potential technical feasibility has been demonstrated. Now the task remains to coordinate the facts that will allow us to develop a system of SCP production that can be justified in economic and biological terms.

RECYCLING ANIMAL WASTES

Only 21.8 percent of the total animal waste nitrogen is identified as amino acids, whereas the remaining 78.2 percent comes from non-protein nitrogen compounds. The average distribution of nitrogen in animal waste products from livestock and poultry is amino acid nitrogen, 21.8 percent; urea nitrogen, 31.6 percent; ammonia nitrogen, 1.3 percent; uric acid nitrogen, 7.2 percent; and other nitrogen compounds, 38.1 percent. From these analyses, it is obvious that animal waste nitrogen that has not been altered is best adapted for utilization by ruminants than by monogastric animals. An important consideration in the utilization of animal waste urea nitrogen is that about 90 percent of the 31.6 million kg produced daily from livestock is excreted in the urine. If urea is to be saved for recycling, facilities must be designed appropriately for recovery. Droppings from caged poultry, especially layers, appear to be most suitable for recycling to ruminants, based on nutritive value and freedom from medicants which often occur in other wastes. New methods must be developed for the conversion of animal waste nitrogen into proteins of high biological value for livestock feeding.

LIST OF CONTRIBUTORS

J. M. ASPLUND, Department of Animal Husbandry, University of Missouri, Columbia, Missouri

W. M. BEESON, Department of Animal Sciences, Purdue University, Lafayette, Indiana

E. M. BICKOFF, Western Regional Research Laboratory, Agricultural Research Service, USDA, Albany, California

JOSEPH CHRISMAN, Western Regional Research Laboratory, Agricultural Research Service, USDA, Albany, California

D. M. DOTY, Fats and Protein Research Foundation, Des Plaines, Illinois

K. J. FREY, Department of Agronomy, Iowa State University, Ames, Iowa

G. O. KOHLER, Western Regional Research Laboratory, Agricultural Research Service, USDA, Albany, California

JEROME H. MANER, Centro Internacional de Agricultura Tropical, Cali, Colombia

JAMES E. OLDFIELD, Department of Animal Science, Oregon State University, Corvallis, Oregon

WILLIAM H. PFANDER, Department of Animal Husbandry, University of Missouri, Columbia, Missouri

K. J. SMITH, National Cottonseed Products Association, Memphis, Tennessee

L. W. SMITH, Animal Science Research Division, Agricultural Research Service, USDA, Beltsville, Maryland

M. A. STEINBERG, Pacific Fisheries Products Technology Center, National Marine Fisheries Service, NOAA, Seattle, Washington

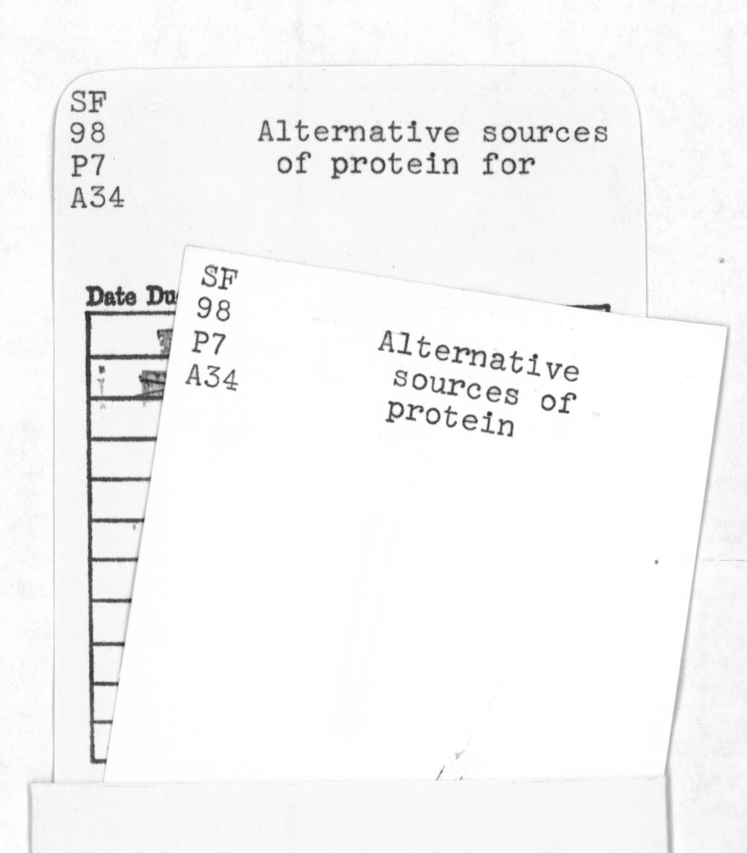

SF
98
P7
A34
Alternative sources
of protein for
Date Du
SF
98
P7
A34
Alternative
sources of
protein

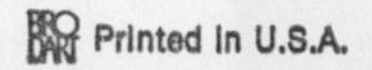

Printed in U.S.A.